Passing Through Nature to Eternity

ProtoCosmos, HyperCosmos, Unified SuperStandard Theory

Stephen Blaha Ph. D.
Blaha Research

Pingree-Hill Publishing
MMXXII

ISBN: 978-1-7372640-5-7

Rev. 00/00/01 May 30, 2022

To Theodore

Some Other Books by Stephen Blaha

All the Megaverse! Starships Exploring the Endless Universes of the Cosmos using the Baryonic Force (Blaha Research, Auburn, NH, 2014)

SuperCivilizations: Civilizations as Superorganisms (McMann-Fisher Publishing, Auburn, NH, 2010)

All the Universe! Faster Than Light Tachyon Quark Starships & Particle Accelerators with the LHC as a Prototype Starship Drive Scientific Edition (Pingree-Hill Publishing, Auburn, NH, 2011).

Unification of God Theory and Unified SuperStandard Model THIRD EDITION (Pingree Hill Publishing, Auburn, NH, 2018).

The Exact QED Calculation of the Fine Structure Constant Implies ALL 4D Universes have the Same Physics/Life Prospects (Pingree Hill Publishing, Auburn, NH, 2019).

Unified SuperStandard Theory and the SuperUniverse Model: The Foundation of Science (Pingree Hill Publishing, Auburn, NH, 2018).

Quaternion Unified SuperStandard Theory (The QUeST) and Megaverse Octonion SuperStandard Theory (MOST) (Pingree Hill Publishing, Auburn, NH, 2020).

Unified SuperStandard Theories for Quaternion Universes & The Octonion Megaverse (Pingree Hill Publishing, Auburn, NH, 2020).

The Essence of Eternity: Quaternion & Octonion SuperStandard Theories (Pingree Hill Publishing, Auburn, NH, 2020).

From Octonion Cosmology to the Unified SuperStandard Theory of Particles (Pingree Hill Publishing, Auburn, NH, 2020).

Beyond Octonion Cosmology (Pingree Hill Publishing, Auburn, NH, 2021).

Integration of General Relativity and Quantum Theory: Octonion Cosmology, GiFT, Creation/Annihilation Spaces CASe, Reduction of Spaces to a Few Fermions and Symmetries in Fundamental Frames (Pingree Hill Publishing, Auburn, NH, 2021).

Available on Amazon.com, bn.com Amazon.co.uk and other international web sites as well as at better bookstores (through Ingram Distributors).

CONTENTS

FIGURES and TABLES

Introduction

It little matters that we cannot see all of Creation. For we have the ability through thought, and sensible consideration, to guide ourselves to a view of the totality of the Cosmos. Mankind has been engaged in creating views of Creation for over 2,500 years. The views have changed as our knowledge of Physics changed. Now we have a new view; a HyperCosmos generated from a ProtoCosmos that seems to give a coherent view of Reality based on Physical considerations.

This book is based on creation and annihilation, as it should be, since these processes are at the base of everything. It takes the author's HyperCosmos (HyperComplex Cosmology) and pursues a deeper analysis, extension, of its features by embodying it in a ProtoCosmos predecessor that leads to it.

A ProtoCosmos theory is developed here based on a three dimension "pre-space" containing a relativistic hydrogen-like atom that generates a geometric spectrum yielding the spaces of the HyperCosmos. (This analysis also suggests conventional calculations of the hydrogen energy spectrum are incomplete for large coupling constants—leading to questions about non-relativistic Charmonium calculations.)

A ProtoCosmos Quantum Electrodynamics is defined with fermions, photons, particle charges and universe interactions.

The book also shows that a parallel study (using GiFT and CASe) of HyperCosmos spaces based on creation and annihilation operators also yields the HyperCosmos set of 10 spaces. The set of spaces is generalized to include the equivalent of positron-like spaces giving anti-universes. A new type of space wave function is defined that embodies universes and anti-universes.

Extending the space concept to particles leads to a new view of confinement with hadrons composed of quark "inner universes" and a new, more sophisticated, type of Bag model.

The set of 256 fundamental fermions in our QUeST space is shown to be reduced to 16 fermions using a generalized General Relativistic transformation to a non-static Fundamental Reference Frame with similar results for the Megaverse and higher spaces.

The book further suggests a basis for the forms of fermions and internal symmetries based on space-time.

It shows how to create fractional dimensions, particles and symmetries in detail. At the infinitesimal Gold Dust level, dynamics becomes classical. The implications of the dynamics for this New Frontier within individual particles include a classical dynamics, a topology, and dust confinement.

Lastly, symmetry splitting of spaces to generate internal symmetries is explored. Thus the book's HyperCosmos explosion of new features from the ultra-large HyperCosmos to the ultra-smsll within particles!

1. Why the HyperCosmos?

Humanity shrinks when confronted with an enormous unknown. Yet one may hope that the human mind, aided by reason, will penetrate the mists and arrive at understanding. In that spirit I have spent several years attempting to comprehend the Cosmos knowing that there is little to guide. This book and the previous two years of books record my progress that has now reached backward to the ProtoCosmos and forward to a deeper HyperCosmos—founded on "gold dust" and consisting of ten spaces that contain internal particle symmetries and space-times. Our universe is the "third" of those spaces as our world is the third in our solar system.

1.1 Origin

The HyperCosmos (Hypercomplex Cosmology) originated in the author's Unified SuperStandard Theory (UST) that was developed over the years preceding 2020. The UST encompasses the known Standard Model, and added a fourth generation of fermions and two additional U(4) symmetry groups (based on conservation laws) that resulted in an expansion of the fundamental fermion spectrum to four layers of four generations. The fermion spectrum totaled to 256 fermions.

In late 2019 the author considered the possibility of hypercomplex numbers for dimensions with a view towards unifying internal symmetries and space-time. Remarkably a theory was found, QUeST, that provided the UST fermion spectrum and internal symmetries united with space-time symmetries.[1] Also a plausible theory of the Megaverse (Multiverse) emerged with six space-time dimensions. Seeing the trend of numbers matching Cayley numbers the HyperCosmos theory of ten physical spaces eventually emerged.

The basis of the HyperCosmos was found to be the creation and annihilation operators of fermion wave functions.[2] A further study of the possibility of a precursor theory led to a Hydrogen-like model whose energy spectrum matched the HyperCosmos spaces spectrum. This theory, called the ProtoCosmos, was set in a three space-time dimension space with one time coordinate using a Weyl unified theory[3] of Gravitation and an Electrodynamics.

The combination of the ProtoCosmos and the consequent HyperCosmos is a complete theory[4] that leads to the known features of the Standard Model and well beyond.

---

[1] Unifying internal symmetries and space-times in an acceptable manner was a primary motivation.

[2] A preliminary theory of the author called Octonion Cosmology was found to be unsatisfactory on grounds of structure and simplicity.

[3] The problems of Weyl's theory that led to its demise were based on local shifts in the light spectrum. This objection is not relevant in the ProtoCosmos space-time.

[4] The author is solely responsible for the ideas and development of these theories, and accepts sole responsibility for them.

1.2 The HyperCosmos Spaces Spectrum

The HyperCosmos spectrum of ten physical[5] spaces is depicted in Fig. 1.1.

THE HyperCosmos SPACES SPECTRUM

Blaha Space Number $N = o_s$	Cayley-Dickson Number n	Cayley Number d_c	Dimension Array d_{dN}	Space-time-Dimension r	CASe Group $su(2^{r/2}, 2^{r/2})$ CASe
0	10	1024	2048 × 2048	18	su(512,512)
1	9	512	1024 × 1024	16	su(256,256)
2	8	256	512 ×512	14	su(128,128)
3	7	128	256 × 256	12	su(64,64)
4	6	64	128 × 128	10	su(32,32)
5	5	32	64 × 64	8	su(16,16)
6	4	16	32 × 32	6	su(8,8)
7	**3**	**8**	**16 ×16**	**4**	**su(4,4)**
8	**2**	4	8 × 8	2	su(2,2)
9	**1**	2	4 × 4	0	su(1,1)

Figure 1.1. The HyperCosmos' ten physical spaces spectrum. The space for our universe, is Blaha number 7,(or Cayley-Dickson number 3) and corresponds to octonions. It is in bold type. Space 6 is the space for the Megaverse (Multiverse).

1.3 Experimental Evidence for the HyperCosmos?

It is unlikely that we will see more than a glimpse of the Megaverse. It is impossible to think that the higher spaces will be detectable astronomically. However, in analogy with the existence question for quarks, we can look for regularities that might reflect the structure of the HyperCosmos.

The most likely circumstantial evidence (currently) is relic fossil evidence: the form of the fundamental fermion spectrum, and the form of the splittings (breakdowns) of the internal symmetry groups in *our* universe.

The key relic fossil evidence is based on the appearance of magic numbers: 4, 8, 16, 32, and 64 in the fermion spectrum structure (See Fig. 15.1 and 15.2. Note four generations.), and in the internal symmetries structure (See Fig. 15.3. Note the splitting of fermions into four's.). Structural evidence will be found in other figures as well. The relic evidence clearly points to QUeST and thus HyperCosmos structuring irrespective of symmetry breaking details.

[5] There is an infinity of additional spaces of physical interest that are not "physical" spaces in that they do not have space-times of interest for universes such as our universe and Megaverses.

HyperCosmos Relics in The Standard Model

1. Quark – lepton quartets; Their combination into octets; Four generations totaling 32 fermions.

2. Symmetries grouped in 8's: by counting real-valued fundamental representation dimensions: strong SU(4) or SU(3)⊗U(1); and SU(2)⊗U(1)⊗O(1, 3); and a U(4) Generation group and a U(4) Layer group.

1.4 What Agent of Implementation?

The elegant mathematical form of the ProtoCosmos and HyperCosmos structure leads to the question of the "Cause" or Source of the structuring and its elegant simplicity. The answer to this question is not known within the current context of Physics. One could suppose an emergent underlayer of unknown dynamics. The physical aspects of this dynamics seem elusive—uniting, as it must, different levels of the total Cosmos.

1.5 Full HyperCosmos Spaces Spectrum

The physical HyperCosmos spectrum of Fig. 1.1, which we derived in Blaha (2022a) and (2022b), suffices for the description of Cosmology in the large. However, a desire for the simplest and most fundamental of theories, as well as an extension of the spectrum to fractionally dimensioned arrays lead to a deeper source for the Cosmos, the ProtoCosmos.

The ProtoCosmos theory developed here suggests negative energy levels leading to anti-spaces and anti-universes.

The appearance of accumulation points in the Hydrogen-like energy spectrum leads to the extension of the spaces spectrum to an infinity of fractionally dimensioned spaces.

The result is the full spectrum of Fig. 1.2.

THE HYPERCOSMOS SPACES SPECTRUM

	Blaha Space Number $N = o_s$	Cayley-Dickson Number n	Cayley Number d_c	Dimension Array column length cd_d	Dimension Array Size d_{dN}	Space-time-Dimension r	CASe Group $su(2^{r/2},2^{r/2})$ CASe
	0	10	1024	2048	2048^2	18	su(512,512)
	1	9	512	1024	1024^2	16	su(256,256)
	2	8	256	512	512^2	14	su(128,128)
	3	7	128	256	256^2	12	su(64,64)
	4	6	64	128	128^2	10	su(32,32)
	5	5	32	64	64^2	8	su(16,16)
	6	4	16	32	32^2	6	su(8,8)
	7	**3**	**8**	**16**	$\mathbf{16^2}$	**4**	**su(4,4)**
	8	2	4	8	8^2	2	su(2,2)
	9	1	2	4	4^2	0	su(1,1)
EXTENSION:							
	10	0	1	2	2^2	-2	U(1)
	11	-1	½	1	1^2	-4	U(½)
	12	-2	¼	½	$½^2$	-6	•
	13	-3	1/8	¼	$¼^2$	-8	•
	14	-4	1/16	1/8	$1/8^2$	-10	•

•
•
•

Figure 1.2. The Extended HyperCosmos space spectrum corresponding to positive energy Hydrogen-like atoms. The negative energy spectrum for anti-spaces (not shown here) has an identical form with negative charged universes. See chapter 6.

2. ProtoCosmos – Genesis of the HyperCosmos

2.1 A ProtoCosmos Parent of the HyperCosmos

The HyperCosmos offers a highly satisfactory spectrum of spaces that can be used to generate universes such as our universe and higher dimension universes such as the Megaverse (Multiverse).

The variety and range of the space spectrum lead us to consider the possibility of a ProtoCosmos that physically generates the HyperCosmos. We call this prior Cosmos, the ProtoCosmos.

The HyperCosmos has a set of spaces that follow a consistent geometric pattern. In this chapter we will create the set of spaces. We will define a three dimension physical model of an atom: an electron-like particle in orbit around a central nucleus. We will start with a space described by H. Weyl's unified theory of gravitation and electrodynamics. This Weyl theory should have a three dimension space-time. Noting the spinors of a three dimension theory are the same as those of a two dimension theory we will use two dimension spinors for one time and one space coordinate.

Locally the atom can be described by 3D Special Relativity and so we use a Hydrogen-like Dirac equation to calculate energy levels. The energy levels constitute a sequence of numbers that we will identify with the dimensions of HyperCosmos spaces.

2.2 The Definition of Spaces

The term space has a variety of definitions. We choose to define space within the framework of Physics as a definition specifying *only* the number of dimensions in the space. A space has no contents. A space has no size. A space is a theoretical construct (a definition).

If an entity is created from matter and energy for a dimension d space then it acquires a size[6] based on its contents with the number of dimensions specified by d. The size can be viewed as delimited by the extent of the mass and energy within it.

The entity may be called a space, although this choice might lead to confusion. Or it could be called a universe. We will use these terms more or less interchangeably.

Given the existence of the mass and energy used in the definition of a universe, the created universe must lie within another universe.[7] The created universe must have a size within the "parent" universe.

We make these observations because the model that we construct generates a spectrum of dimensions in the form of an energy spectrum. This spectrum will coincide with the dimensions of the HyperCosmos spectrum of spaces by construction. Thereby the HyperCosmos is defined by the ProtoCosmos.

---

[6] We expect the initial size to be a point that subsequently expands. See Blaha (2004) or (2021d).

[7] We do not allow "naked" mass and energy outside of universes.

2.3 A Bottom Up View of the Spectrum of Spaces

The spectrum of spaces of Fig. 1.2 invites the view that it corresponds to an inverted spectrum of states where d_{cd} plays a role analogous to (negative) energy, and the Blaha space number O_s plays the role of an energy level index. Note the energy accumulation point at zero as $N = O_s \rightarrow \infty$. Note also the geometric scaling behavior of the energy d_{cd}. Fig. 2.1 shows the inverted spectrum plotted as an energy spectrum.

$-d_{dN}$ Column Size	Space-time Dimension r	Blaha Space Number N = O_s
0	$-\infty$	∞
$\bullet\bullet\bullet$		
$-1/16^2$	-12	15
$-1/8^2$	-10	14
$-1/4^2$	-8	13
$-1/2^2$	-6	12
-1^2	-4	11
-2^2	-2	10
-4^2	0	9
$\bullet\bullet\bullet$		
-1024^2	16	1
-2048^2	18	0

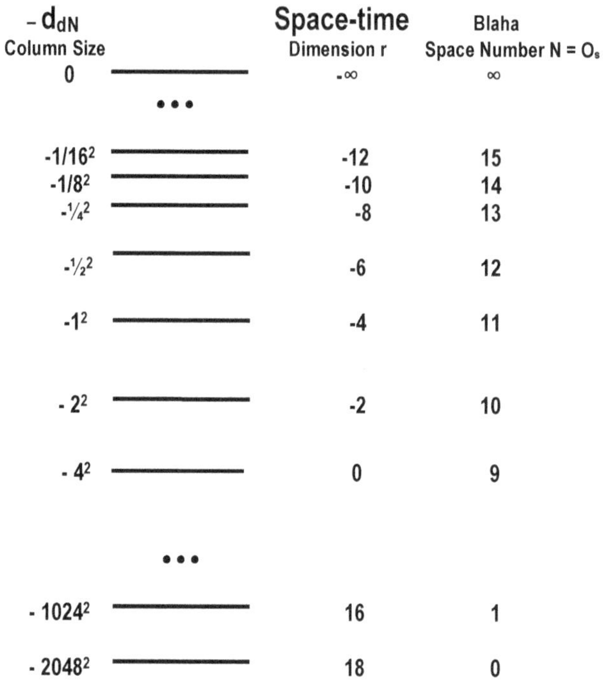

Figure 2.1. Extended HyperCosmos spectrum plotted as an Inverted energy spectrum.

2.4 The ProtoCosmos

We assume a "space-time" exists in the ProtoCosmos, within which a dynamics can take place that generates the HyperCosmos. We rule out 0 and 1 dimension spaces since they preclude dynamics as we usually know it.

We might have assumed a two-dimension "space-time" exists with one space and one time dimension—each with real-valued coordinates. We choose[8] a three dimension space-time since its Dirac equation leads to the same energy spectrum. Three dimensions supports a Weyl-like unified theory. This choice is the simplest choice of space-time supporting a conventional dynamics.

---

[8] We chose three dimensions to have an angular momentum defined in the space. It results in a better match with the Case four dimension calculation seen below.

2.4.1 Kaluza-Klein Theories and Weyl's Unified Gravitation and Electromagnetism Theory

Alternately, it is also reasonable to assume a third dimension to support a "unified" theory of General Relativity and "Electromagnetism" in a manner similar to the Kaluza-Klein[9] unified theory.

However the author believes that Weyl's unified theory[10] of gravitation and electromagnetism is a better choice. Weyl based his theory on an extension of Riemannian geometry to include a path dependence for the transfer of lengths between points in space. In Weyl's theory length is path dependent and a length at one point is different at other points. His theory derives General Relativity and Electrodynamics in four dimensions. The theory aroused great interest around 1920. But it was discarded after it was realized[11] that spectral lines for a substance would differ noticeably from point to point in space. Weyl sought to get around the issue by asserting the theory's lengths were not related to *real* lengths measured by instruments thus making the lengths and spectral lines predicted by the theory unobservable. The theory thus does not apply in our space-time.

The situation differs in the ProtoCosmos. We see that there is no direct connection to our universe. Thus questions of spectral lines, and so on, are not relevant. So we assume Weyl's theory applies in a three dimension ProtoCosmos, which is effectively in two dimensions: one time and one spatial dimension.

We assume one center of force exists (the nucleus or "Cosmic Egg") and use a Weyl-like unified theory for gravitation and electrodynamics in two dimensions. The result is a *minimal* theory in two dimensions of a Hydrogen-like atom that will be used to generate the dimensions of the HyperCosmos spectrum of Spaces.. Thereby, we achieve our goal of a minimalist (ProtoCosmos) model.

2.4.2 Hydrogen-like Atom Model

We can assume a flat space-time with two dimensions[12] where Special Relativity holds locally around the nucleus. We then assume a hydrogen-like atom model with an electron-like fermion, a center of force (nucleus[13]), and a "Weylian" electro-magnetic like interaction. The nucleus will be seen later to have a large charge Ze and thus a large coupling constant Zα. (The value of α is different from the four dimension case.)

We further assume the electromagnetic interaction has a 1/x form[14] generated from a Lagrangian term:

$$\mathcal{L} = - \frac{1}{2} m_A^2 A^\mu A_\mu$$

---

[99] See Appelquist, T.; Chodos, T.; Freund, P. G. O.; *Modern Kaluza-Klein Theories* (Addison-Wesley, Menlo Park, CA, 1987) and references therein.

[10] H. Weyl, *Space-Time-Matter* (1920, Dover Publications, New York, 1952). See also W. Pauli, *Theory of Relativity* (1921, Pergamon Press, London, 1952).

[11] See Pauli's discussion. op. cit. *Theory of Relativity*, p. 1.96.

[12] The two dimension case differs inconsequentially from the three dimension case.

[13] One might view this as a "Cosmic Egg."

[14] To make contact with the Case solution of the Dirac equation with a 1/r potential. See section 2.5.3 and 2.5.4.

where m_A is a mass-like parameter corresponding to A^μ, which is massless. The dynamic equation for A^μ is[15]

$$m_A{}^2 A^\mu = 0$$

If we restrict x to $0 \leq x \leq \infty$, and use

$$\int_0^\infty dk\, e^{ikx} = i/(x + i\varepsilon)$$

then

$$A^0 = V(x) = 1/(2\pi\, m_A{}^2) \int_0^\infty dk\, e^{-ikx} = 1/(\, m_A{}^2 x)$$

giving a potential of the same form as the radial r potential of K. M. Case[16] in four dimensions. The electromagnetic propagator has the form

$$S^{\nu\mu}(x - y) = <0|T(A^\nu(x)A^\mu(y))|0> = 1/(2\pi)^2 \int d^2k\, e^{-ik\cdot(x-y)}\, \eta^{\nu\mu}/(m_A{}^2)$$

where $\eta^{\nu\mu} = \text{diag}(1, -1)$ is the metric. Note

$$m_A{}^2\, S^{\nu\mu}(x - y) = \eta^{\nu\mu}\, \delta^2(x - y)$$

The 2-dimension Dirac equation is:

$$(\beta\gamma^1 p + \beta m + V(x))\psi = H\psi = E\psi \tag{2.1}$$

where m is the fermion mass and

$$\beta = \begin{bmatrix} 1 & 0 \\ 0 & -1 \end{bmatrix} \tag{2.2}$$

$$\gamma^1 = \begin{bmatrix} 0 & 1 \\ 1 & 0 \end{bmatrix} \tag{2.3}$$

$$p = \hbar/i\ d/dx \tag{2.4}$$

$$V(x) = -Z\alpha/x \tag{2.5}$$

[15] Note that gauge invariance is absent.
[16] K. M. Case, Phys Rev, **80**, 797 (1950).

for a central charge Z.[17] Using the above Dirac equation formulation we obtain the same radial equations as Case for four dimensions. After some algebra the upper (G) and lower (F) components of ψ satisfy[18,3]

$$dF/dx - \kappa F/x = (m - E - Z\alpha/x)G \tag{2.6}$$
$$dG/dx + \kappa G/x = (m + E + Z\alpha/x)F \tag{2.7}$$

where we set the angular momentum $l = 0$ *to obtain the two-space-time dimension case giving:*

$$\kappa = -1 \text{ for } j = ½ \quad \text{and} \quad \kappa = 0 \text{ for } j = -½ \qquad (2/8)$$

The standard solution for the two-dimension energy eigenvalues is the same as for the four-dimension case:[19]

$$E_N = m[1 + (Z\alpha/(N - (j + ½) + ((j + ½)^2 - Z^2\alpha^2)^{½})^2]^{-½} \tag{2.9}$$

Case found a different expression for E_N which corresponds to eq. 2.9 for small $Z\alpha$

$$E_N \approx m[1 - ½\, Z^2\alpha^2/N^2] \tag{2.10}$$

but differs for large $Z\alpha$ as we show below.
The known large $Z\alpha$ problems of the standard solution eq. 2.9 are

Oscillatory solutions
E_N has negative values

These problems are known to be due to the x → 0 anomalous behavior of the wave function ψ. They are resolved using Case's expression for the energy E_N as shown below.

2.5 Case's Calculation of E_N

Case calculated quantum mechanical wave functions for singular potentials in four dimensions. His relativistic four dimension solution for the hydrogen atom is the same as that of the two dimension case above if the angular momentum is set to zero: $l = 0$. He calculated the singular potential energy levels based on requiring a fixed phase for wave functions at the origin. This approach guarantees orthogonal eigensolutions.

Case found a different form for the energy eigenvalues in four dimensions, which applies to our two dimension case as well. It appears that the small $Z\alpha$ form of E_N does correspond for both calculations. The large $Z\alpha$ limit differs. Case's limiting E_N does not have the difficulties of eq. 2.9.

---

[17] We absorb an $m_A{}^2$ factor in $Z\alpha$.
[18] C . G. Darwin, Proc. Roy. Soc (London) **A118**, 654 (1928). W. Gordon, Z. Physik, **48**, 11 (1928).
[19] H. A. Bethe and E. E. Salpeter, *Quantum Mechanics of One and Two-electron Atom,s* Academic Press Inc., New York 1957. Bjorken (1964).

We present the calculations of the energy eigenvalues for large and small Zα in some detail below since there are subtleties. The calculations apply for both four and two dimensions.

2.5.1 Small Zα vs. Large Zα

Eqs. 2.6 and 2.7 combine to yield a 1/x behavior and a $1/x^2$ behavior resulting in a singular effective potential.

$$V_{eff} = E\delta/x + (Z^2\alpha^2 - \chi^2)/x^2 \tag{2.11}$$

when expressed in terms of the two dimension coordinate x. δ and χ are constants defined by Case. (We absorb the $m_A{}^2$ factor in Zα.)

It appears that the 1/x behavior governs the form of E_N for large x while the $1/x^2$ behavior governs the form of E_N for small x near the origin. Fig. 2.2 shows the controlling effects of the effective potential.

It further appears that the small Zα values of E_N is sensitive to the large x behavior of 1/x, and the large Zα values of E_N behavior is sensitive to the small x behavior of $1/x^2$. Thus the difference between the conventional E_N of eq. 2.9 and Case's expression in eq. 2.10 below.

This discussion applies equally to the two and four dimension cases.

Small Zα	Large Zα >> 1
Sensitive to large x behavior of V 1/x	Sensitive to small x behavior of V $1/x^2$
Determines E_N of Darwin and Others	Determines E_N of Case Case E_N also gives small Zα form of Darwin and Others

Figure 2.2. Effect of terms in V_{eff} on the behavior.in.the computation.

2.5.2 Case's Form of E_N

Case specifies the Dirac energy levels using his procedure as

$$(B' - N\pi)/\lambda - [\tan^{-1} \lambda/(\chi - \delta(m + E))]/\lambda - \arg[\Gamma(-EZ\alpha/(m^2 - E^2)^{1/2} + i\lambda)]/\lambda = \ln(m^2 - E^2)^{1/2} \tag{2.12}$$

where $E = E_N$ and where

$$\begin{aligned} \chi &= 0 \qquad \text{for } j = -1/2 \\ \chi &= -1 \qquad \text{for } j = 1/2 \\ \lambda &= (Z^2\alpha^2 - \chi^2)^{1/2} \\ \delta &= Z\alpha/(m^2 - E^2)^{1/2} \end{aligned} \tag{2.13}$$

2.5.3 The Darwin and Others Form of E_N for Zα << 1

For small Zα and χ = 0 we find

2.5.3.1 Third Term of Eq. 2.12

$$\arg[\Gamma(-EZ\alpha/(m^2 - E^2)^{½} + i\lambda)] \approx 0$$

using $$E_N \approx m[1 - ½\, Z^2\alpha^2/N^2 + \varepsilon]$$

where ε represents higher order terms in Zα, we find

$$(m^2 - E_N{}^2)^{½} \approx (m^2Z^2\alpha^2/N^2 + \ldots)^{½}$$
$$\approx m\, Z\alpha/N + \ldots$$
$$E_NZ\alpha/(m^2 - E^2)^{½} \approx N + \ldots$$

Therefore for large N

$$\arg[\Gamma(-EZ\alpha/(m^2 - E^2)^{½} + i\lambda)] \approx \arg[\Gamma(-N - \ldots + i\, Z\alpha)] \approx 0$$

since the imaginary part is negligible. (Note: … represents higher order terms in Zα in for E_N.)

2.5.3.2 Right Side of Eq. 2.12

We now express eq. 2.12 as

$$(B' - N\pi) - [\tan^{-1} \lambda/(\chi - \delta(m + E))] - \arg[\Gamma(-EZ\alpha/(m^2 - E^2)^{½} + i\lambda)] = \lambda\ln(m^2 - E^2)^{½} \quad (2.12')$$

We find

$$\lambda\ln(m^2 - E^2)^{½} \approx 0$$

since

$$Z\alpha \ln(Z\alpha) \rightarrow 0$$

as Zα becomes small.

2.5.3.3 The Resulting Eigenvalue Equation

We now have the eigenvalue condition form:

$$B' - N\pi \approx \tan^{-1} (-2m\lambda/\delta) \quad (2.12'')$$

We now suggest the constant λ appearing in eq. 2.12″ is should possibly be replaced it with 1.[20] This assumption leads to the known small Zα result fot E_N:

$$E_N \approx m[1 - ½\, Z^2\alpha^2/N^2] \quad (2.10)$$

Otherwise agreement is not possible. Thus we have

$$B' - N\pi + \tan^{-1}(1/(\delta 2m)) \approx 0$$

---

[20] Note if λ were in absolute value $\lambda = |Z^2\alpha^2 - \chi^2|^{½}$ then $\lambda \approx 1$ for $j = ½$ and small Zα.

with

$$\delta = Z\alpha/(m^2 - E^2)^{½}$$

Using the identity

$$\text{atan}(1/x) = \pi/2 - \text{atan}(x) \qquad \text{for } x > 0$$

we find

$$\tan^{-1}(1/(\delta 2m)) = \pi/2 - \tan^{-1}(\delta 2m)$$
$$\approx \pi/2 - 2m\delta = \pi/2 - 2m\, Z\alpha/(m^2 - E^2)^{½}$$

with the result

$$B' - N\pi + \pi/2 - 2m\, Z\alpha/(m^2 - E^2)^{½} \approx 0$$

Setting $B' + N\pi = 0$ we obtain

$$E^2 = m^2 - 4m^2 Z^2\alpha^2/(N\pi)^2 \tag{2.14}$$

and

$$E \approx m - ½\, m\, Z^2\alpha^2/N^2$$

if

$$\alpha = e^2/8$$

and

$$\alpha_E{}^2 = 4\alpha^2/\pi^2$$

where e is the charge of the electron and $\alpha_E = e^2/4\pi$. Early Physics did not standardize the form of α.

We thus have shown the Case form of E_N has the proper small Zα limiting behavior. *We suggest, subject to further study, the correct eigenvalue equation is*

$$B' - N\pi - \tan^{-1}[1/(\chi - \delta(m + E))] - \arg[\Gamma(-EZ\alpha/(m^2 - E^2)^{½} + i\lambda)] = \lambda \ln(m^2 - E^2)^{½} \tag{2.12'''}$$

with λ replaced by one in the second term. The above sections are for 4 and 2 space-time dimensions for the Zα << 1 Case eigenvalue condition. Note that the eigenvalue condition eq. 2.14 has a negative energy solution. This solution has a negative energy electrons (of negative charge like positive energy electrons) binding to form an atom. It is correctly discarded since the Dirac electron sea prevents an atom with a negative energy electron. A complete quantum field theory calculation, which includes the filled vacuum, would eliminate these states.

2.5.4 Case Form of E_N for Zα >> 1

We now consider the ProtoCosmos Dirac equation eigenvalue condition, eq. 2.12‴, in two space-time dimensions for Zα >> 1. Evaluating the terms for Zα >> 1 we find the approximate equation:[21]

$$B' - N\pi \approx \lambda \ln(m^2 - E^2)^{½} \tag{2.15}$$

[21] The second and third terms in eq. 2.12″ would at most modify the arbitrary constant B′ without any significant consequences.

with the result

$$E^2 = m^2[\,1 - \exp((2B'' - 2\pi N)/\lambda)] \qquad (2.16)$$

with m = exp(B′ – B″). The energy spectrum is geometric:

$$E_N = m[\,1 - \exp((2B'' - 2\pi N)/\lambda)]^{1/2} \qquad (2.17)$$

$$\cong m[1 - \tfrac{1}{2}\exp(2B''/\lambda)\,(\exp(\pi/\lambda)^{-2N}] \qquad (2.18)$$

Eq. 2.18 is similar to the eigenvalue spectrum calculated using the Schrödinger equation in Blaha (2022b) except for relativistic effects.

The relativistic Dirac equation limits the most negative value of E_N:

$$1 \geq \exp((2B'' - 2\pi N)/\lambda)$$

or

$$N \geq -B''/\pi \qquad (2.19)$$

At $N = -B''/\pi$, $E_N = 0$. For smaller N the solution for E_N^2 (eq. 2.16) would be negative and E_N would be imaginary.

N can be large without limit. As $N \to \infty$ we see

$$E_N \to m \qquad (2.20)$$

an accumulation point. Thus the positive branch of E_N lies between 0 and m: $0 \leq E_N <$ m. We make contact with the HyperCosmos by defining the spectrum with

$$E_N \cong m - \tfrac{1}{2} m \exp(2B''/\lambda) \exp(\pi/\lambda)^{-2N} \qquad (2.21)$$
$$= m + E_{SN}$$

where

$$E_{SN} = E_S' d_{dN}'$$

with the energy factor

$$E_S' = -\tfrac{1}{2} m \exp(2B''/\lambda) \qquad (2.22)$$

and

$$d_{dN}' = \exp(\pi/\lambda)^{-2N}$$

We now map the number of degrees of freedom to the spectrum of the HyperCosmos in Fig. 2.1 by setting

$$\exp(\pi/\lambda) = 2 \qquad (2.23)$$

which imples

$$\lambda = 4.532 \cong Z\alpha$$

verifying the assumption that $Z\alpha \gg 1$, and suggesting extreme localization of the energy levels.

We now let

$$d_{dN}' = 2^{-2N} = d_{dN}2^{-22} \tag{2.24}$$

where

$$d_{dN} = 2^{22-2N} \tag{2.25}$$

Then we set $E_0 = ½ m$ by letting

$$-½ m \exp(2B''/\lambda) = -½ m \tag{2.26}$$

Defining

$$E_S'' = 2^{-22}E_S' = -½ m\, 2^{-22} \tag{2.27}$$

we see

$$E_{SN} = E_S''d_{dN} \tag{2.28}$$
$$= -2^{-23}md_{dN}$$

and thus

$$E_N \cong m - 2^{-23}md_{dN} = m - ½ m\, 2^{-2N} \tag{2.29}$$

where $d_{d0} = 2^{22} = 2048^2$ for Blaha space number N = 0 as in Fig. 2.3.

Thus

$$d_{dN} = 2^{23}(1 - E_N/m) \tag{2.29a}$$

Thus the inverted spectrum of E_N and d_{dN} in Fig. 2.3 is obtained. We have thus established the connection between the two space-time dimension Dirac equation of the ProtoCosmos for a Hydrogen-like atom and the dimensions of the spectrum of spaces of the HyperCosmos.[22] The corresponding derivation of the HyperCosmos spaces' spectrum from spaces of creation/annihilation operators appears in Chapter 3.

Note we have specified the lowest energy E_0 to correspond to the Blaha number N = 0 space in the ten spaces HyperCosmos spectrum. See Fig. 2.1.

2.6 Spaces Maps to Internal Space-Times and HyperCosmos Spectrum Bounds

In chapter 4 we consider the relation of N to the space-time dimension r within each space by considering the creation/annihilation operators group structure of fermions in each space. We find

$$N = Os = ½(18 - r) \tag{2.30}$$

where r is the space-time dimension. Then

$$d_{dN} = 2^{22-2N} = 2^{r+4} \tag{2.31}$$

The lowest energy E_0 specifies the uppermost HyperCosmos space. The lowest physical space-time dimension r = 0 limits the physical spaces spectrum to $N \le 9$. The 10 space spectrum of the HyperCosmos results.

The relativistic solution sets upper and lower bounds on the energy spectrum unlike the non-relativistic example solution in Blaha (2022b).

---

[22] Blaha (2022b) derives the spectrum similarly in a non-relativistic Schrödinger hydrogen-like atom model.

2.7 The Positive Energy Spectrum Implied by the Model

The upper bound of the spectrum is the accumulation point:

$$E_N = m \tag{2.30}$$

at $N = \infty$.

The low energy bound is set by noting $E_N = -m < 0$ if $N = -1$ thus violating the assumption of a positive E_N branch in eq. 2.17. Thus E_0 is the lowest positive energy as indicated by Fig. 2.3.

	E_N	d_{dN}	Space-time Dimension r	Blaha Space Number $N = O_s$	Entropy S
Positive Energy					
	$m = m - ½ m\, 2^{-\infty}$	0	$-\infty$	∞	0
		•••			
	$m - ½ m\, 2^{-30}$	$1/16^2$	-12	15	0
	$m - ½ m\, 2^{-28}$	$1/8^2$	-10	14	0
	$m - ½ m\, 2^{-26}$	$¼^2$	-8	13	0
	$m - ½ m\, 2^{-24}$	$½^2$	-6	12	0
	$m - ½ m\, 2^{-22}$	1^2	-4	11	0
	$m - ½ m\, 2^{-20}$	2^2	-2	10	$2\, M\, k_{BN} \ln 2$
		•••			•••
	$m - ½ m\, 2^{-2}$	1024^2	16	1	$20\, M\, k_{BN} \ln 2$
	$m - ½ m = m/2$	2048^2	18	0	$22\, M\, k_{BN} \ln 2$
	0				
Negative Energy					
	$-m + ½ m = -m/2$	2048^2	18	0	$22\, M\, k_{BN} \ln 2$
	$-m + ½ m\, 2^{-2}$	1024^2	16	1	$20\, M\, k_{BN} \ln 2$
		•••			•••
	$-m + ½ m\, 2^{-20}$	2^2	-2	10	$2\, M\, k_{BN} \ln 2$
	$-m + ½ m\, 2^{-22}$	1^2	-4	11	0
	$-m + ½ m\, 2^{-24}$	$½^2$	-6	12	0
	$-m + ½ m\, 2^{-26}$	$¼^2$	-8	13	0
	$-m + ½ m\, 2^{-28}$	$1/8^2$	-10	14	0
	$-m + ½ m\, 2^{-30}$	$1/16^2$	-12	15	0
		•••			
	$-m = -m + ½ m\, 2^{-\infty}$	0	$-\infty$	∞	0

Figure 2.3. The energy spectrum of two dimension Dirac model of hydrogen-like atom. The dimension based entropy is shown for each HyperCosmos space.

2.8 From Energy Eigenvalues to the HyperCosmos Spectrum of Spaces

The model atom that we have constructed generates a spectrum of dimensions in the form of an energy spectrum. This spectrum agrees with the dimensions of the HyperCosmos spectrum of spaces.

In chapter 3 we suggest that the "electron" of the ProtoCosmos atom contains a universe of one of the HyperCosmos spaces. We suggest that it is at the lowest energy level initially. Thus it is a universe corresponding to the Blaha number N = 0 HyperCosmos space.

The energy levels (principal quantum numbers) of the bound state wave functions specify the wave function solution just like the four dimension Hydrogen bound state wave functions are determined by its principal quantum number (and thus the energy level). The ProtoCosmos wave function has both two dimension coordinates x, and r dimension space (universe) coordinates u. See eq. 2.41 below. The x coordinate behavior is determined by the binding. The r coordinates are free of the binding. They specify the universe. The bound state wave function selects the dimension of the space based on the atomic level's principal quantum number.[23]

This assignment is part of a Quantum Electrodynamics that we suggest for the ProtoCosmos in chapter 3. The equivalent of electrons is charged space fermion particles. The interaction is electrodynamic-like with ProtoCosmos photons. The theory is a three dimension unified Weyl-like theory of Gravitation and Electrodynamics.

In chapter 6 we show universes have an electric-like charge. Thus we view the ProtoCosmos Hydrogen-like atom as similar to a Hydrogen atom. The ProtoCosmos atom generates an energy level spectrum equal to the HyperCosmos spaces' spectrum. One could envision a ProtoCosmos "photon" striking the atom and boosting it to a higher energy level analogous to a different HyperCosmos space universe.

Thus the HyperCosmos spectrum of spaces can be defined by the ProtoCosmos Hydrogen-like atom model. The dimensions of HyperCosmos spaces then determine their overall nature.

The creation/annihilation operator groups described in chapter 4 also yield the HyperCosmos spectrum. They "fix" the space-time part of each HyperCosmos space. As a result the HyperCosmos structure is fully determined.

2.9 Negative Energies and Anti-Spaces

There are, of course, negative energy solutions as we mentioned in section 2.4.3. For four dimension hydrogen atoms we ruled out these solutions since negative energy

---

[23] We have chosen to match the principal quantum number of the atom with the Blaha number of the HyperCosmos space (universe), in the bound state wave function. This choice is arbitrary in the sense that any space might be specified. We view the situation as analogous to angular momentum in a 4 dimension Hydrogen atom. There, the dynamic equation has a momentum term that specifies the angular momentum l for the wave function as the Casimir operator eigenvalue $l(l+1)$. In the present case there is no term in the Dirac equation fixing the universe's Blaha number N. We choose to specify it by fiat. A more complete theory of the ProtoCosmos atom could specify the number N universe it in terms of a Casimir operator for the $SU(2^{21-2N})$ group, which has 2^{22-2N} real-valued dimensions in its fundamental representation. See chapter 15.

electrons could not exist in the filled Dirac sea. For our two dimension hydrogen-like atom we can assume a filled Dirac sea also. However a filled sea may not be present.

In the case of the $Z\alpha \gg 1$ spectrum there is a set of negative energy eigensolutions for E_N:

$$E_N^2 = m^2[\,1 - \exp((2B'' - 2\pi N)/\lambda)] \tag{2.31}$$

with, the negative energy levels (Fig. 2.3)

$$E_N = -\,m[\,1 - \exp((2B'' - 2\pi N)/\lambda)]^{1/2} \tag{2.32}$$

$$\cong -\,m[1 + (-\tfrac{1}{2}\exp(2B''/\lambda))\,(\exp(\pi/\lambda)^{-2N}] \tag{2.33}$$

Letting

$$E_S'^{-} = \tfrac{1}{2}\,m\,\exp(2B''/\lambda) \tag{2.34}$$

we are led to

$$\begin{aligned} E_{SN}^{-} &= E_S''^{-}\,d_{dN} \\ &= 2^{-23} m d_{dN} \end{aligned} \tag{2.35}$$

$$E_N^{-} \cong -\,m + E_{SN}^{-} \tag{2.36}$$

Negative energy levels are limited by $-\tfrac{1}{2}$ m and $-$ m in our hydrogen-like atom. See Fig. 2.3.

2.10 Universe Holes – Anti-Spaces and Anti-Universes

The presence of negative energy states of the hydrogen-like atom raises the possibility of "universe holes"—hole states for anti-universes just as there are positive charge hole states, positrons, for electrons. We can imagine a process creating a universe–anti-universe pair.

Naturally one will ask: How does an anti-universe differ from a universe? At first glance it would seem reasonable (given our level of understanding of universes) to believe universes begin with "pure" matter and energy, and anti-universes begin with "pure" antimatter. Then as they evolve universes develop an anti-matter component due to (very weak) interactions, and anti-universes develop a matter component due to (very weak) interactions. This concept could account for the preponderance of matter in our universe.

However in our study of space wave functions later (Chapter 6) we will see it is equally reasonable to view universes and anti-universes as the same except for a "charge", which is positive for universes and negative for anti-universes. This view preserves an analogy with the similarity of electrons and positrons except for charge. The charge of universes is then relevant in a parent space for electrodynamic-like interactions between universes.

2.11 Extension of the HyperCosmos Spectrum to Anti-Spaces

If we follow the guidance of the hydrogen-like atom model energy levels then the HyperCosmos spectrum of spaces must be extended to contain both a spaces spectrum and an anti-spaces spectrum of the same structure. The anti-spaces spectrum is used to define anti-universes.

When we define wave functions for universes and anti-universes then pair creation and annihilation must be considered. The definition of second quantized wave functions necessarily involves creation and annihilation operators and the specification of inter-universe interactions.

2.12 ProtoCosmos Free Fermion Wave Functions

Based on the energy spectrum and consequent spaces dimensions we can define free wave functions for spaces in two dimension space using the PseudoQuantum formalism. We begin by defining creation and annihilation operators for spaces (universes):

$$b_{iMN\alpha}(k, q)^\dagger \quad \text{and} \quad b_{iMN\alpha}(k, q) \tag{2.37}$$

for i = 1, 2 where M is the mass-energy of the created universe, N is the Blaha number, $r = 18 - 2N$ (by eq. 2.30)) is the number of space-time dimensions of the universe with momentum $q = (q_1, q_2, \ldots, q_r)$, with $q_r = M$, and the index α ranges from 1 to $d_{dN} - r$ where $d_{dN} = 2^{22-2N}$. We take u_r to be the time variable within the universe. The index α labels non-space-time components for the $d_{dN} - r$ dimension internal symmetry group representation for the universe.

The momentum k is for the two dimension space. The momentum q is an r-vector for the embedded space (universe). We similarly define hole creation and annihilation operators for anti-spaces (anti-universes)

$$d_{iMN\alpha}(k, q)^\dagger \quad \text{and} \quad d_{iMN\alpha}(k, q) \tag{2.38}$$

The non-zero anticommutator relations are

$$\{b_{iMN\alpha}(k, q), b_{jMN\alpha'}(k', q')^\dagger\} = \delta_{\alpha\alpha'}\delta(k - k')\, \delta^{r-1}(q - q') \tag{2.39}$$
$$\{d_{iMN\alpha}(k, q), d_{jMN\alpha'}(k', q')^\dagger\} = \delta_{\alpha\alpha'}\delta(k - k')\, \delta^{r-1}(q - q') \tag{2.40}$$

for i, j = 1, 2 with $i \neq j$.

The PseudoQuantum wave functions are:

$$\psi_{iMN\alpha}(\mathbf{x}, t, \mathbf{u}, t_u) = \int dk \int d^{r-1}q\, \mathfrak{N}(q, k)[b_{iMN\alpha}(k, q)e^{-i(k\cdot x + q\cdot u)} + \\ + d_{iMN\alpha}(k, q)^\dagger e^{i(k\cdot x + +q\cdot u)}] \tag{2.41}$$

for i = 1, 2 where $\mathfrak{N}$ is a normalization constant, where m is the fermion mass in the two dimension space and M is the mass-energy in the universe space of Blaha number N and r is the universe space-time dimension. We assume the $k^2 = m^2$ of the fermion in the space and $q^2 = M^2$ in the N = 0 universe space. Note $x = (x^0, \mathbf{x})$, $u = (u^0, \mathbf{u})$, $t = x^0$, and $t_u = u^0$.

The $b_{iMN\alpha}(k, q)^\dagger$ operators generate electron-like states of positive charge. The $d_{iMN\alpha}(k, q)^\dagger$ operators generate "hole" states (positron equivalents) of negative charge.

The wave function equal time anti-commutator is

$$\{\psi_{iMN\alpha}(\mathbf{x}, t, \mathbf{u}, t_u), \psi_{jMN\alpha'}(\mathbf{x}', t, \mathbf{u}', t_u)^\dagger\gamma^0\} = \int dk \int d^{r-1}q \int dk' \int d^{r-1}q'\ \mathfrak{N}(q, k)\ \mathfrak{N}(q', k')\ \{[b_{iMN\alpha}(k, q)e^{-i(k\cdot x+q\cdot u)} +$$

$$+\ d_{iMN\alpha}(k, q)^\dagger e^{i(k\cdot x + +q\cdot u)}],\ [b_{jN\alpha'}(k', q')^\dagger e^{i(k'\cdot x' + q'\cdot u')} + d_{jMN\alpha'}(k', q')e^{-i(k'\cdot x' + q'\cdot u')}]\ \gamma^0\}$$

$$= \int dk \int d^{r-1}q \int dk' \int d^{r-1}q'\ \mathfrak{N}(q, k)\ \mathfrak{N}(q', k')\delta_{\alpha\alpha'}\delta(k - k')\ \delta^{r-1}(q - q')\cdot$$
$$\cdot[e^{-i(k\cdot x+q\cdot u)}e^{i(k'\cdot x' + q'\cdot u')} + e^{-i(k\cdot x+q\cdot u)}e^{-i(k'\cdot x' + q'\cdot u')}]$$

$$= \delta_{\alpha\alpha'}\int dk \int d^{r-1}q\ \mathfrak{N}(q, k)^2\ [e^{-i(k\cdot x+q\cdot u)}e^{i(k\cdot x' + q\cdot u')} + e^{i(k\cdot x+q\cdot u)}e^{-i(k\cdot x' + q\cdot u')}]$$

for $i \neq j$ where the space's time is t, the independent embedded universe's time is t_u and where the 1×1 matrix $\gamma^0 = (1)$. Note the spatial parts are in bold typeface. Note also $t \neq t_u$. They are independent variables.

If $\mathfrak{N}(q, k)^2 = (2\pi)^{-r/2}$, then

$$\{\psi_{iMN\alpha}(\mathbf{x}, t, \mathbf{u}, t_u), \psi_{jMN\alpha'}(\mathbf{x}', t, \mathbf{u}', t_u)^\dagger\} = \delta_{\alpha\alpha'}\delta(\mathbf{x} - \mathbf{x}')\ \delta^{r-1}(\mathbf{u} - \mathbf{u}') \qquad (2.42)$$

A one universe state is created from the two dimension vacuum by

$$|MN\alpha kq> = b_{2MN\alpha}{}^\dagger(k, q)|0> \qquad (2.43)$$

using the vacuum $|0>_2$ of PseudoQuantum theory. The universe initial wave function is

$$\psi_{MN\alpha}(x, u, k\ , q) = <x, u|MN\alpha kq> \qquad (2.44)$$

Note that a superposition of wave functions using a distribution of momenta, and mass M introduces a quantum aspect in the initial universe particle state.

Interaction terms such as the electromagnetic-like interaction for atoms given earlier can be defined.

$\psi_{MN\alpha}(x, u)$ forms a representation of $su(2^{9-N}, 2^{9-N})$ if one includes both internal symmetries and the universe's space-time. See Figs. 1.1 and 1.2.

2.13 The Entropy of a HyperCosmos Universe

Entropy is usually not thought to be dependent on the dimensions of a system since dimensions are not thought to vary. In the case of the spaces of the HyperCosmos spectrum dimensions vary from space to space.

We can view the dimensions of a space with dimension d_{dN} as specifying a set of microstates (in principle) described by a mass-energy M. Entropy then arises in the creation of a universe (in analogy with chemical reactions.)

The energy of a microstate per dimension can be viewed as M. We suggest the entropy of the creation process of a universe or anti-universe is

$$S_N = k_{BN} \ln(d_{dN})^M \tag{2.45}$$

where k_{BN} is Boltzmann's constant for space N. Using

$$d_{dN} = 2^{22-2N} \tag{2.46}$$

we find

$$S_N = M\, k_{BN} \ln d_{dN}$$
$$S_N = (22 - 2N)\, M\, k_{BN} \ln 2 \tag{2.47}$$

Spaces with N > 11 have negative S_N and are thus precluded since absolute zero corresponds to S = 0.

Fig. 2.3 shows larger dimension spaces have larger entropy as one expects since entropy depends on the number of microstates. More dimensions implies more microstates.

2.14 An Entropic Direction of Time

Since entropy increases with time it appears possible to use the entropy in a universe (space) to specify a time variable. This time variable could be assumed to be an implicit time variable in a non-static Fundamental Reference Frame (Space Rest Frame). As a result General Relativistic transformations to static reference frames may be better defined.

2.15 Revisit Charmonium?

Charmonium, bound states of charmed and anti-charmed quarks, was investigated in 1974 within the framework of non-relativistic Quantum Mechanics model based on the assumption that the mass of the quarks would justify a non-relativistic theory.[24] This work assumed a potential

$$V(r) = -\kappa/r + r/a^2 \tag{2.48}$$

where $\kappa = 0.61$ and a = 2.38 GeV^{-1} and the charmed quark mass was 1.84 GeV. It has been noted that the 1/r term had a large coupling constant relative to the linear potential r term. The largeness of the 1/r coupling, coupled with the analysis of the dependence of the energy eigenvalues on the form of the potential for small and large r, suggests that the Charmonium model should have been constructed in a relativistic Dirac equation model based on our conclusions in section 2.5.1 above. There we found:

> It appears that the 1/x behavior governs the form of E_N for large x while the $1/x^2$ behavior governs the form of E_N for small x near the origin. Fig. 2.2 shows the controlling effects of the effective potential.
>
> It further appears that the small $Z\alpha$ values of E_N is sensitive to the large x behavior of 1/x, and the large $Z\alpha$ values of E_N behavior is sensitive to the small x behavior of $1/x^2$. Thus the difference between the conventional E_N of eq. 2.9 and Case's expression in eq. 2.10.

---

[24] See E. J. Eichten et al, arXiv:hep-ph 0206018 (2012) for references for this model and more recent papers.

This discussion applies equally to the two and four dimension cases.

Small $Z\alpha$	Large $Z\alpha >> 1$
Sensitive to large x behavior of V	Sensitive to small x behavior of V
$1/x$	$1/x^2$
Determines E_N of Darwin and Others	Determines E_N of Case
	Case E_N also gives small $Z\alpha$ form of Darwin and Others

where x is the radial coordinate r. The relativistic Dirac equation has an implicit $1/r^2$ behavior due to the $1/r$ electromagnetic potential near r = 0. This behavior may be embedded in the large $1/r$ coupling in the non-relativistic model. The Case calculation described above has a large coupling constant form for the energy, which differs significantly from the small coupling constant form.

Thus Charmonium might best be modeled within a relativistic Dirac equation, as Case did for electromagnetism, with a $1/r$ term with a large coupling constant plus a linear term for quark confinement. The effect of the linear term may only appear in the energy spectrum's fine structure.

3. HyperCosmos Generation from a Hydrogen-like Atom in ProtoCosmos QED

The Hydrogen-like atom considered in chapter 2 has a central nucleus (the Cosmic Egg) and an electron-like particle circling it in an orbit. See Fig. 3.1. We now consider the possibility that the circling particle is a space fermion particle as described in section 2.12 with Blaha number N = 0 containing a Blaha number N = 0 space (universe) of the HyperCosmos. This universe is of the largest size with dimension 2048^2.

Note the fermion wave function $\psi_{iMN\alpha}(x, u)$ in section 2.12 satisfies the Dirac equation in the ProtoCosmos with the universe part playing the role of a "mute" witness. (See chapter 6 for more on space fermions.) The Dirac equation is independent of the index α and the internal coordinates u of the embedded HyperCosmos universe.

3.1 ProtoCosmos QED

In section 2.12 and in chapter 6 we show that space fermion wave functions have charged electron-like states and charged positron-like states. The charge operator has the conventional form $\Sigma\ (b^\dagger b - d^\dagger d)$. There is an electromagnetic analogue that is used to create a Hydrogen-like atom with a space fermion "circling" a nucleus in energy levels similar to the HyperCosmos spectrum of spaces. See Chapter 14 for charge operators, and anti-universes.

The interaction is electrodynamics-like with ProtoCosmos photons. One could envision a ProtoCosmos "photon" striking an atom boosting it to a higher energy level, which is analogous to a different HyperCosmos space universe.

Space fermions could interact via ProtoCosmos space photon exchange.

Thus we can have a QED-like theory in a three (or two) dimension unified Weyl theory of Gravitation and Electrodynamics.

3.2 HyperCosmos Spaces Generation

The energy levels of the Hydrogen-like bound state wave functions specify the wave function solution just like the four dimension Hydrogen bound state wave functions are determined by its principal quantum number (and thus its energy level).

A ProtoCosmos wave function has both two dimension coordinates x, and r dimension space (universe) coordinates u. See eq. 2.41. The x coordinate behavior is determined by the binding. The r space-time coordinates are free of the binding. They specify the universe. The bound state wave function selects the dimension of the space based on the ProtoCosmos electron's atomic level principal quantum number.

Within the N = 0 electron's universe in the ground state atom will be space fermions that can annihilate to generate an N = 1 space (universe) as described in our earlier work (and later in this book). Space fermions in the N = 1 HyperSpace may annihilate to produce an N = 2 universes. This process may iterate to generate a sequence of universes for N = 3, … , 9 thus populating the HyperCosmos spectrum of spaces (universes). See Fig. 3.1.

The other principal quantum number wave functions each have other embedded HyperCosmos spaces within them. The entire HyperCosmos spectrum corresponds, on a 1:1 basis, with the HyperCosmos spaces' spectrum.

We thus see the generation of the HyperCosmos universes from the Cosmic Egg atom with a circling space fermion in three (two) dimension space-time. It contains the parent universe of all universes of the HyperCosmos.

The ProtoCosmos is the progenitor of the HyperCosmos. The ProtoCosmos is united with the HyperCosmos spaces, which, in turn, are united with QUeST and the Unified SuperStandard Theory in a theory that encompasses all known interactions including internal symmetries and Quantum Gravitation: A union of Cosmology and Elementary Particle Physics.

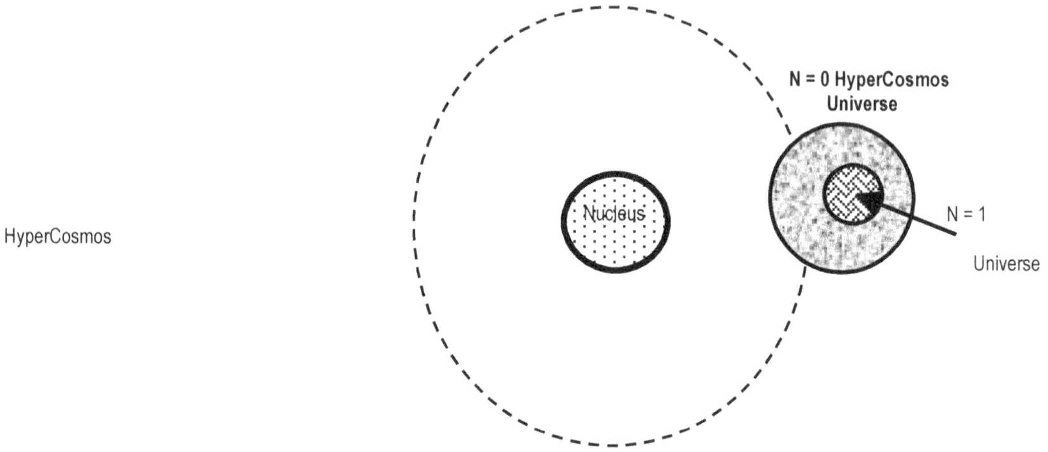

Figure 3.1. A ProtoCosmos atom with a circling space fermion in the N = 0 energy level. It contains an N = 0 HyperCosmos universe with an embedded N = 1 universe. The N = 1 universe may contain a nested series of universes of type N = 2, 3, … , 9.

4. Derivation of the HyperCosmos Spaces from Creation/Annihilation Operators

4.1 GiFT for Particle Creation/Annihilation Operator Transformations

In Blaha (2021i) and (2021j) we developed the features of Generalized Field Theory (GiFT).[25] In this chapter we describe it and apply its formalism to the cases of bosons and fermions. *We will show the HyperCosmos spectrum of spaces naturally emerges in agreement with the derivation in chapters 2 and 3.*

We begin by considering two 4-dimension coordinate systems denoted α and β and develop a relation between the quantizations in these coordinate systems for scalar bosons, and for fermions where the spins of the fermions in the respective coordinate systems are the same. Then we consider the fermion case where the energies and spins of states in one coordinate system correspond to superpositions of energy and spin states in other coordinate system.

Since higher dimension space-times appear in the HyperCosmos we develop the relation between quantizations of coordinate systems in higher dimensions.

In the discussions of creation/annihilation operators we will determine features for a given momentum, usually denoted k. The results will typically lead to CASe groups[26] that are infinite tensor products over all values of k in each HyperCosmos space.

Note on PseudoQuantum Theory

Some advantages of PseudoQuantum Field Theory are:

1. For Quantization in any coordinate system in flat or curved space-times with an invariant definition of asymptotic particle states. *This is especially important for the higher dimension spaces of the HyperCosmos.* An n particle asymptotic state in one coordinate system has a unitarily equivalent n particle asymptotic state in any other coordinate system. Therefore particle number is invariant under change of coordinate system. This is important for the HyperCosmos and the Unified SuperStandard Theory in curved space-times. *It is also important for quantization in higher dimensional spaces such as those of the HyperCosmos.* The method was developed in the late 1970's by the author to provide a quantization procedure which supports a unique particle interpretation of states in arbitrary non-static space-times where a different, or no, global time-like coordinate (Killing vector) exists. PseudoQuantum Field Theory, which we developed in a series of papers and books,[27] also can be formulated in the HyperCosmos

[25] GiFT is the Quantum Field Theory based on PseudoQuantum Theory and Two-Tier Theory—both created by this author. PseudoQuantum Field Theory is needed in HyperCosmos Physics. It is needed for proper quantization in arbitrary coordinate systems in higher dimension HyperCosmos spaces. See Blaha (2021i) for details.

[26] See Blaha (2021i) and (2021j) for CASe details.

[27] See Blaha (2017b) for the discussion of the PseudoQuantum field theory formalism for Higgs particles in our Extended Standard Model. See chapter 20 of Blaha (2017b), and earlier books, for a more detailed view than that presented here.

Spectrum of spaces. For example, we can use it to implement the Higgs Mechanism to generate particle masses and symmetry breaking.

2. PseudoQuantum Field Theory enables one to define Higgs particle dynamics in such a way that a non-zero vacuum expectation value cleanly separates from the quantum field part of the Higgs fields. This technique can be used in symmetry breaking mechanisms, mass generation, and possible generation of coupling constants as vacuum expectation values.

3. It supports the canonical definition of higher derivative field theories through the use of the Ostrogradski bootstrap. We have used it to construct a fourth order theory of the Strong interaction that has color confinement and a linear r potential. The potential part of this theory was used by the Cornell group to calculate the Charmonium spectrum. (See Blaha (2017b) for details.)

Scalar Boson GiFT and CASe Creation/Annihilation Transformations

4.1.1 GiFT for Scalar Creation/Annihilation Operator Transformations

We illustrate the procedure by considering the case of a scalar particle in four dimensions[28] and expand on the development given in section 4 of S. Blaha, "The Local Definition of Asymptotic Particle States", IL Nuovo Cimento **49A**, 35 (1979), which we call I. *The PseudoQuantum Field Theory in I, and other1970s Blaha papers listed in Appendix A, is essential for the development of satisfactory GiFT and CASe formulations.* See Blaha (2021i) and (2021j) for additional details.

We associate a scalar particle with two scalar fields: $\varphi_1(x)$ and $\varphi_2(x)$. We choose $\varphi_1(x)$ to have a zero equal time commutator with $d\varphi_1(x)/dx^0$ and $\varphi_2(x)$ to have a conventional equal time commutator with $d\varphi_2(x)/dx^0$. Conceptually $\varphi_1(x)$ is a "classical" field and $\varphi_2(x)$ is a quantum field. A Lagrangian that implements these choices of commutation relations is:

$$\mathcal{L} = \partial^\mu \varphi_1(x)\partial_\mu\varphi_2(x) - \tfrac{1}{2}\, \partial^\mu \varphi_1(x)\partial_\mu\varphi_1(x) - m^2\, \varphi_1(x)\varphi_2(x) + \tfrac{1}{2}\, m^2\, \varphi_1(x)^2 \tag{4.1}$$

$$(\Box + m^2)\varphi_1(x) = 0 \tag{4.2}$$

$$(\Box + m^2)\varphi_2(x) - (\Box + m^2)\varphi_1(x) \quad = 0 \tag{4.3}$$

The canonical momenta are

$$\pi_1 = d\varphi_2(x)/dt - d\varphi_1(x)/dt \tag{4.4}$$

$$\pi_2 = d\varphi_1(x)/dt \tag{4.5}$$

and the equal time commutation relations are

$$[\varphi_i(x), \pi_j(y)] = i\delta_{ij}\delta(\mathbf{x} - \mathbf{y}) \tag{4.6}$$

$$[\varphi_i(x), \varphi_j(y)] = [\pi_i(x), \pi_j(y)] = 0 \tag{4.7}$$

for i, j = 1, 2 implying

---

[28] The case of a space with a higher dimension spacetime is completely analogous. See Appendix A for a description of Blaha's PseudoQuantum Theory and Two-Tier Quantum Theory.

$$[\varphi_1(x), d\varphi_1(y)/dt] = 0 \tag{4.8}$$
$$[\varphi_2(x), d\varphi_2(y)/dt] = i\delta^3(\mathbf{x} - \mathbf{y}) \tag{4.9}$$
$$[\varphi_1(x), d\varphi_2(y)/dt] = i\delta^3(\mathbf{x} - \mathbf{y}) \tag{4.10}$$

The *most* general mode expansion of the fields in the coordinate system A with Fourier momentum α is

$$\varphi_{1A}(x) = \Sigma_\alpha\, [a_{1\alpha}\, f_\alpha(x) + a^\dagger_{1\alpha}\, f_\alpha{}^*(x)] \tag{4.11}$$
$$\varphi_{2A}(x) = \Sigma_\alpha\, [a_{2\alpha} f_\alpha(x) + a^\dagger_{2\alpha}\, f_\alpha{}^*(x)] \tag{4.12}$$

and in coordinate system B with Fourier momentum β is

$$\varphi_{1B}(x) = \Sigma_\beta\, [a_{1\beta}\, f_\beta(x) + a^\dagger_{1\beta}\, f_\beta{}^*(x)] \tag{4.13}$$
$$\varphi_{2B}(x) = \Sigma_\beta\, [a_{2\beta}\, f_\beta(x) + a^\dagger_{2\beta}\, f_\beta{}^*(x)] \tag{4.14}$$

The creation/annihilation operators of A have the commutation relations:

$$[a_{i\alpha'}, a_\alpha] = 0 \tag{4.15}$$
$$[a^\dagger_{i\alpha'}, a^\dagger_{j\alpha}] = 0$$
$$[a_{i\alpha}, a^\dagger_{j\alpha'}] = (1 - \delta_{ij})\delta^3(\boldsymbol{\alpha} - \alpha') \tag{4.16}$$

for i, j = 1, 2. Two related vacuums are defined: $|0>_1$ and $|0>_2$ by

$$a_{1\alpha}|0>_{2\alpha} = a^\dagger_{1\alpha}|0>_{2\alpha} = 0 \tag{4.17}$$
$$a_{2\alpha}|0>_{1\alpha} = a^\dagger_{2\alpha}|0>_{1\alpha} = 0$$
$$a_{i\alpha}|0>_{i\alpha} \neq 0 \qquad a^\dagger_{i\alpha}|0>_{i\alpha} \neq 0 \tag{4.18}$$

for i = 1, 2 and all α and α′. And similarly for the creation/annihilation operators of B with α replaced with β.

The general coordinate system B Fourier expansion in terms of A creation/annihilation operators is

$$\varphi_{1B}(x) = \Sigma_\beta\, \Sigma_\alpha\, [(c_{11}a_{1\alpha} + c_{12}a_{2\alpha} + C_{11}a^\dagger_{1\alpha} + C_{12}a^\dagger_{2\alpha})f_\beta(x) + \\ + (c'_{11}a^\dagger_{1\alpha} + c'_{12}a^\dagger_{2\alpha} + C'_{11}a_{1\alpha} + C'_{12}a_{2\alpha})f_\beta{}^*(x)] \tag{4.19}$$

$$\varphi_{2B}(x) = \Sigma_\beta\, \Sigma_\alpha\, [(c_{21}a_{1\alpha} + c_{22}a_{2\alpha} + C_{21}a^\dagger_{1\alpha} + C_{22}a^\dagger_{2\alpha})\, f_\beta(x) + \\ + (c'_{21}a^\dagger_{1\alpha} + c'_{22}a^\dagger_{2\alpha} + C'_{21}a_{1\alpha} + C'_{22}a_{2\alpha})f_\beta{}^*(x)] \tag{4.20}$$

where the c_{ij} and C_{ij}, and c'_{ij} and C'_{ij} are all functions of α.

We define sets of Bogoliubov transformations B_1 and B_2 that give

$$a_{i\beta} = \Sigma_\alpha\, (f_\beta, f_\alpha)B_{2\alpha}B_{1\alpha}a_{i\alpha}(B_{1\alpha}B_{2\alpha})^{-1} = \Sigma_\alpha\, (f_\beta, f_\alpha)[c_{i1}a_{1\alpha} + c_{i2}a_{2\alpha} + C_{i1}a^\dagger_{1\alpha} + C_{i2}a^\dagger_{2\alpha}] \tag{4.21}$$

$$a^\dagger_{i\beta} = \Sigma_\alpha\, (f^*_\beta, f^*_\alpha)\, B_{2\alpha}B_{1\alpha}a^\dagger_{i\alpha}(B_{1\alpha}B_{2\alpha})^{-1} = \Sigma_\alpha\, (f^*_\beta, f^*_\alpha)[c'_{i1}a^\dagger_{1\alpha} + c'_{i2}a^\dagger_{2\alpha} + C'_{i1}a_{1\alpha} + C'_{i2}a_{2\alpha}] \tag{4.22}$$

4.1.1.1 The B_1 Transformation

The B_1 transformation has the form:

$$B_{1\alpha}(x_1, x_2) = \exp[x_1\Gamma^1{}_{3\alpha}]\exp[x_2\Gamma^1{}_{2\,\alpha}] \tag{4.23}$$

where the Hermitian operators $\Gamma^1{}_{1i\alpha}$ are

$$\Gamma^1{}_{3\alpha} = (a^\dagger{}_{2\alpha}a_{1\alpha} + a_{2\alpha}a^\dagger{}_{1\alpha})/2 \tag{4.24}$$
$$\Gamma^1{}_{2\alpha} = i(a^\dagger{}_{2\alpha}a^\dagger{}_{1\alpha}\ - a_{2\alpha}a_{1\alpha})/2 \tag{4.25}$$

We also define

$$\Gamma^1{}_{1\alpha} = -\ (a^\dagger{}_{2\alpha}a^\dagger{}_{1\alpha}\ + a_{2\alpha}a_{1\alpha})/2 \tag{4.26}$$

and

$$\Gamma^1{}_{4\alpha} = i(a^\dagger{}_{2\alpha}a_{1\alpha} - a_{2\alpha}a^\dagger{}_{1\alpha})/2 \tag{4.27}$$

Together the three operators satisfy su(1,1) algebra commutation relations:

$$[\Gamma^1{}_{1\alpha}, \Gamma^1{}_{2\alpha'}] = -i\delta_{\alpha\alpha'}\,\Gamma^1{}_{3\alpha} \qquad [\Gamma^1{}_{2\alpha}, \Gamma^1{}_{3\alpha'}] = i\delta_{\alpha\alpha'}\,\Gamma^1{}_{1\alpha} \qquad [\Gamma^1{}_{3\alpha}, \Gamma^1{}_{1\alpha'}] = i\delta_{\alpha\alpha'}\,\Gamma^1{}_{2\alpha} \tag{4.28}$$

They transform $a_{i\alpha}$ and $a^\dagger{}_{i\alpha}$ to terms of the form $c_{i\alpha}a_{i\alpha} + c_{i\alpha}'a^\dagger{}_{i\alpha}$ where $c_{i\alpha}$ and $c_{i\alpha}'$ are constants.

Some properties of the su(1,1) algebra are detailed in S. Blaha, "The Local Definition of Asymptotic Particle States", IL Nuovo Cimento **49A**, 35 (1979).

4.1.1.2 The B_2 Transformation

The B_2 transformation has the form:

$$B_{2\alpha}(x_1, x_2) = \exp[x_1\Gamma^2{}_{3\alpha}]\exp[x_2\Gamma^2{}_{2\,\alpha}] \tag{4.29}$$

where the Hermitian operators $\Gamma^2{}_{1i\alpha}$ are

$$\Gamma^2{}_{3\alpha} = (a^\dagger{}_{1\alpha}a_{1\alpha} + a_{1\alpha}a^\dagger{}_{1\alpha})/2 \tag{4.30}$$
$$\Gamma^2{}_{2\alpha} = (a^\dagger{}_{2\alpha}a_{2\alpha} + a_{2\alpha}a^\dagger{}_{2\alpha})/2 \tag{4.30}$$
$$\Gamma^2{}_{1\alpha} = (a^\dagger{}_{2\alpha}a^\dagger{}_{2\alpha}\ +\ a_{2\alpha}a_{2\alpha})/2 \tag{4.31}$$

We also define the Hermitian operators

$$\Gamma^2{}_{4\alpha} = (a^\dagger{}_{1\alpha}a^\dagger{}_{1\alpha}\ + a_{1\alpha}a_{1\alpha})/2 \tag{4.32}$$

and

$$\Gamma^2{}_{5\alpha} = i(a^\dagger{}_{2\alpha}a^\dagger{}_{2\alpha} - a_{2\alpha}a_{2\alpha})/2 \tag{4.33}$$
$$\Gamma^2{}_{6\alpha} = i(a^\dagger{}_{1\alpha}a^\dagger{}_{1\alpha} - a_{1\alpha}a_{1\alpha})/2 \tag{4.34}$$

They transform $a_{1\alpha}$ and $a^\dagger_{1\alpha}$ to have additional terms of the form $c_\alpha a_{2\alpha} + c_\alpha' a^\dagger_{2\alpha}$ where c_α and c_α' are constants; and they transform $a_{2\alpha}$ and $a^\dagger_{2\alpha}$ to have additional terms of the form $c_\alpha a_{1\alpha} + c_\alpha' a^\dagger_{1\alpha}$.

The complete set of Γ operators are part of an su(1,1) group. This group appears in Twistor Theory. A relation/unification of the group of these Bogoliubov transformations and Twistor Theory may eventually emerge.

The transformation $B_{2\alpha}B_{1\alpha}$ generates the maps of eqs. 4.21 and 4.22. As we will see next the key terms for the transition between annihilation/creation operators of two coordinate systems, A and B, have the form:

$$[c'_{i1}a^\dagger_{1\alpha} + c'_{i2}a^\dagger_{2\alpha} + C'_{i1}a_{1\alpha} + C'_{i2}a_{2\alpha}] \tag{4.35}$$

if we construct states based on $a^\dagger_{2\beta}$ and $a^\dagger_{2\alpha}$ with the vacuum $|0\rangle_2$. Note there are four constants, which are functions of α, in eq. 4.35. These constants are determined by the four parameters in $B_{2\alpha}B_{1\alpha}$. Thus the transformation between coordinate systems embodied in $B_{2\alpha}B_{1\alpha}$ determines the constants in eq. 4.35.

4.1.2 States in the Respective Coordinate Systems

The B states are superpositions of the A states as eq. 4.21 and 4.22 demonstrate. The one particle state in the B coordinate system is

$$a^\dagger_{2\beta}\,|0\rangle_2 = \Sigma_\alpha\,(f^*_\beta, f^*_\alpha)[c'_{21}a^\dagger_{1\alpha} + c'_{22}a^\dagger_{2\alpha} + C'_{21}a_{1\alpha} + C'_{22}a_{2\alpha}]\,|0\rangle_2$$

$$= \Sigma_\alpha\,(f^*_\beta, f^*_\alpha)[c'_{22}a^\dagger_{2\alpha} + C'_{22}a_{2\alpha}]\,|0\rangle_2 \tag{4.36}$$

where the $a^\dagger_{2\alpha}$ term generates a one particle A state, and the $a_{2\alpha}$ term generates an A particle negative energy state. The two particle B coordinate system state

$$a^\dagger_{2\beta}\,a^\dagger_{2\beta}\,|0\rangle_2$$

generates a superposition in momenta α) of two A particles, two A negative energy particles and an A vacuum state term.

Later, when we consider fermions, we will see that parton momentum distributions, xF(x), in deep inelastic e-p scattering may partially originate in the transition from a rest frame coordinate system to a different internal coordinate system.

4.2 Scalar Particle Creation/Annihilation Space (CASe)

The set of creation and annihilation operators a_1, a_2, $a^\dagger_1$, $a^\dagger_2$ form an operator basis for spaces, for each value α.[29] In the scalar particle case in four space-time dimensions we define a four dimension space that we call the su(1,1) Creation/Annihilation Space or the CASe(1,1) space that supports an su(1,1)

---

[29] **Note this calculation requires the use of PseudoQuantum fields.** Two fields × two operators (an a and $a^\dagger$ per field) = 4 real operators. Thus 4 real-valued coordinates. . The fundamental su(1,1) representation has 4 real-valued coordinates. Match. See eq. 4.28 as well.

representation for each α. We call su(1,1) a GiFT transformation group. Defining basis "ortho" vectors $a_\alpha = (a^\dagger_{1\alpha}, a^\dagger_{2\alpha}, a_{1\alpha}, a_{2\alpha})$ we specify complex coordinates for an irreducible su(1,1):

$$x_\alpha = (x_{1\alpha}{}^0, x_{2\alpha}{}^0, x_\alpha{}^1, x_{2\alpha}{}^1) \tag{4.37}$$

Defining the metric

$$ds^2 = |dx_{1\alpha}{}^0|^2 + |dx_{2\alpha}{}^0|^2 - |dx_{1\alpha}{}^1|^2 - |dx_{2\alpha}{}^1|^2 \tag{4.38}$$

we note that there is an su(1,1) group invariance.

Thus for scalar particles an su(1,1) symmetry exists for each Fourier component α in a space-time of any number of space-time dimensions.

The vectors in the space have the form

$$a_\alpha \cdot x_\alpha = x_{1\alpha}{}^0 a^\dagger_{1\alpha} + x_{2\alpha}{}^0 a^\dagger_{2\alpha} + x_\alpha{}^1 a_{1\alpha} + x_{2\alpha}{}^1 a_{2\alpha} \tag{4.39}$$

The "ortho" vectors a_α form an algebra reminiscent of hypercomplex quaternions. The lack of complete commutativity is not a source for disquiet when one considers the anti-commutativity of quaternion, octonion and sedenion multiplication. It can be viewed positively as a step in the direction of a larger unified theory such as

$$\text{Spinor Space} + \text{CASe(n,m)} \tag{4.40}$$

seen later. CASe is enlarged by spin for fermions in the various space-time dimensions.

See chapter 6 for a discussion of CASe groups for bosons.

Note: A General Relativistic transformation between coordinate systems with Killing vectors may not induce a GiFT transformation. A General Relativistic transformation between a static and non-static coordinate system does induce a GiFT CASe transformation.

Fermion GiFT and CASe Creation/Annihilation Transformations

4.3 GiFT for *Fermion* Creation/Annihilation Operator Transformations

The GiFT formalism developed in Blaha (2021i) and in section 4 of S. Blaha, "The Local Definition of Asymptotic Particle States", IL Nuovo Cimento **49A**, 35 (1979), which we call I, is directly extendable to fermions. We begin by considering 4-dimension space-time coordinate systems. Following the framework developed in section 4 of I we define fermion fields $\psi_1(x)$ and $\psi_2(x)$ in a coordinate system A labeled with α as

$$\psi_{1A}(x) = \Sigma_{\alpha,s}[b_{1\alpha s}u_{\alpha s}f_\alpha(x) + d^\dagger_{1\alpha s}v_{\alpha s}f_\alpha{}^*(x)] \tag{4.41}$$
$$\psi_{2A}(x) = \Sigma_{\alpha,s}[b_{2\alpha s}u_{\alpha s}f_\alpha(x) + d^\dagger_{2\alpha s}v_{\alpha s}f_\alpha{}^*(x)] \tag{4.42}$$

and in another coordinate system B labeled with β as

$$\psi_{1B}(x) = \Sigma_{\beta,s}[b_{1\beta s}u_{\beta s}\, g_\beta(x) + d^\dagger_{1\beta s}v_{\beta s}\, g_\beta{}^*(x)] \qquad (4.43)$$

$$\psi_{2B}(x) = \Sigma_{\beta,s}[b_{2\beta s}u_{\beta s}\, g_\beta\,(x) + d^\dagger_{2\beta s}v_{\beta s}\, g_\beta{}^*(x)] \qquad (4.44)$$

where $f_\alpha(x)$ and $g_\beta(x)$ are Fourier components.

To begin, we assume the spin sectors transform independently. We also anticipate the Fourier expansions are related in a manner. Then for spin value s we have

$$b_{1\beta s} = \Sigma_{\alpha,x}\,(g_\beta,\, f_\alpha)\; u^\dagger_{\beta s}u_{\alpha s}\,(c_{11s}b_{1\alpha s} + c_{12s}b_{2\alpha s} + C_{11s}b^\dagger{}_{1\alpha s} + C_{12s}b^\dagger{}_{2\alpha s}) \qquad (4.45a)$$

$$b_{2\beta s} = \Sigma_{\alpha,x}\,(g_\beta,\, f_\alpha)\; u^\dagger_{\beta s}u_{\alpha s}\,(c_{21s}b_{1\alpha s} + c_{22s}\, b_{2\alpha s} + C_{21s}b^\dagger{}_{1\alpha s} + C_{22s}b^\dagger{}_{2\alpha s}) \qquad (4.45b)$$

$$d^\dagger{}_{1\beta s} = \Sigma_{\alpha,x}\,(g^*{}_\beta\, f^*{}_\alpha)\; v^\dagger{}_{\beta s}v_{\alpha s}\,(c'_{11s}d^\dagger{}_{1\alpha s} + c'_{12s}d^\dagger{}_{2\alpha s} + C'_{11s}d^\dagger{}_{1\alpha s} + C'_{12s}d^\dagger{}_{2\alpha s}) \qquad (4.46a)$$

$$d^\dagger{}_{2\beta s} = \Sigma_{\alpha,x}\,(g^*{}_\beta\, f^*{}_\alpha)\; v^\dagger{}_{\beta s}v_{\alpha s}\,(c'_{21s}d^\dagger{}_{1\alpha s} + c'_{22s}d^\dagger{}_{2\alpha s} + C'_{21s}d^\dagger{}_{1\alpha s} + C'_{22s}d^\dagger{}_{2\alpha s}) \qquad (4.46b)$$

generalizing eq. 118 of I where the c_{ij} are all constants and where x is a space-like surface. The inner products have the form

$$\Sigma_x\,(g_\beta,\, f_\alpha) = i\;\Sigma_x\, g^*{}_\beta \overleftrightarrow{\partial_0} f_\alpha \qquad (4.46c)$$

where the subscript "0" indicates the time-like coordinate. For example, in rectilinear coordinates

$$\Sigma_x\,(g_{k'},\, f_k) = i\;\Sigma_x\, g^*{}_k \overleftrightarrow{\partial_0} f_k = \delta^3(\mathbf{k'} - \mathbf{k}) \qquad (4.46d)$$

for momenta k and k′.

The non-zero anti-commutation relations of the creation/annihilation operators are:

$$\{b_{i\alpha s}(\alpha),\, b^\dagger{}_{j\alpha s'}(\alpha')\} = \{d_{i\alpha s}(\alpha\,),\, d^\dagger{}_{j\alpha s'}(\alpha')\} = (\,1 - \delta_{ij})\delta_{ss'}\; \delta^3(\boldsymbol{\alpha} - \boldsymbol{\alpha}') \qquad (4.47)$$
$$\{b_{i\beta s}(\beta),\, b^\dagger{}_{j\beta s'}(\beta')\} = \{d_{i\beta s}(\beta),\, d^\dagger{}_{j\beta s'}(\,\beta')\} = (\,1 - \delta_{ij})\delta_{ss'}\; \delta^3(\boldsymbol{\beta} - \boldsymbol{\beta}')$$

Note the transformations of eqs. 4.45-46 superimpose "momenta" in relating the coordinate system quantizations. But they do not superimpose spins.[30] As a result the transformations can be factored into two sets: one set for spin up and one set for spin down.

4.3.1 Transformations of the b and b† Operators without Spin Mixing or b – d Mixing

The sets of Bogoliubov transformations B_1 and B_2 for either spin up or down b creation/annihilation operators are defined by

---

[30] We consider spin mixing later.

$$B_{1\alpha}(x_1, x_2) = ex[x_1\Gamma^1{}_{3\alpha}] \exp[x_2\Gamma^1{}_{2\,\alpha}] \qquad (4.48)$$

where the Hermitian operators $\Gamma^1{}_{1i\alpha}$ are

$$\Gamma^1{}_{3\alpha}(b_{1\alpha s}, b_{2\alpha s}) = (b^\dagger{}_{2\alpha s}b_{1\alpha s} + b_{2\alpha s}b^\dagger{}_{1\alpha s})/2 \qquad (4.49)$$
$$\Gamma^1{}_{2\alpha}(b_{1\alpha s}, b_{2\alpha s}) = i(b^\dagger{}_{2\alpha s}b^\dagger{}_{1\alpha s} - b_{2\alpha s}b_{1\alpha s})/2 \qquad (4.50)$$

We also define

$$\Gamma^1{}_{1\alpha}(b_{1\alpha s}, b_{2\alpha s}) = -(b^\dagger{}_{2\alpha s}b^\dagger{}_{1\alpha s} + b_{2\alpha s}b_{1\alpha s})/2 \qquad (4.51)$$

and

$$\Gamma^1{}_{4\alpha}(b_{1\alpha s}, b_{2\alpha s}) = -(b^\dagger{}_{2\alpha s}b_{1\alpha s} + b_{2\alpha s}b^\dagger{}_{1\alpha s})/2 \qquad (4.52)$$

The B_2 transformation has the form:

$$B_{2\alpha}(x_1, x_2) = \exp[x_1\Gamma^2{}_{3\alpha}] \exp[x_2\Gamma^2{}_{2\,\alpha}] \qquad (4.53)$$

where the Hermitian operators $\Gamma^2{}_{1i\alpha}$ are

$$\Gamma^2{}_{3\alpha} = \Gamma^1{}_{3\alpha}(b_{1\alpha s}, b_{1\alpha s}) \qquad (4.54)$$
$$\Gamma^2{}_{2\alpha} = \Gamma^1{}_{3\alpha}(b_{2\alpha s}, b_{2\alpha s}) \qquad (4.55)$$
$$\Gamma^2{}_{1\alpha} = \Gamma^1{}_{1\alpha}(b_{2\alpha s}, b_{2\alpha s}) \qquad (4.56)$$

We also define the Hermitian operators

$$\Gamma^2{}_{4\alpha} = \Gamma^1{}_{1\alpha}(b_{1\alpha s}, b_{1\alpha s}) \qquad (4.57)$$
$$\Gamma^2{}_{5\alpha} = \Gamma^1{}_{2\alpha}(b_{2\alpha s}, b_{2\alpha s}) \qquad (4.58)$$
$$\Gamma^2{}_{6\alpha} = \Gamma^1{}_{2\alpha}(b_{1\alpha s}, b_{1\alpha s}) \qquad (4.59)$$

The above set of operators generate the su(1,1) transformations to the forms of eqs. 4.46. The su(1,1) representation has four real components for b_i, and $b_i^\dagger$ for i = 1, 2.

4.3.2 Transformations of the d and d† Operators without Spin Mixing or b – d Mixing

The operators for "holes" have an analogous form to the b operators. These operators are part of the operators of another su(1, 1).

4.4 Operators of Types b and d *with* Spin Mixing and b – d Mixing

These operators, plus those of section 4.1, are part of the operators of an su(4,4), which contains operators mixing spins, and b and d operators. The su(4, 4) fundamental representation, which mixes spins, and b's and d's, has eight real components for b_i, $b_i^\dagger$ d_i, and $d_i^\dagger$ for i = 1, 2 for each spin.

$$su(2, 2)\otimes su(2, 2) \rightarrow su(4, 4) \qquad (4.60)$$

We can anticipate that certain General Relativistic transformations may cause spin mixing. Some typical su(4, 4) Creation/Annihilation Space (CASe) transformations are (in the notation of section 4.1)

$$B_{1\alpha}(x_1, x_2) = \exp[x_1\Gamma^1{}_{3\alpha}]\exp[x_2\Gamma^1{}_{2\alpha}] \tag{4.61}$$

with

$$\Gamma^1{}_{3\alpha}(b_{1\alpha s}, b_{2\alpha s'})$$
$$\Gamma^1{}_{2\alpha}(b_{1\alpha s}, b_{2\alpha s'})$$
$$\Gamma^1{}_{1\alpha}(b_{1\alpha s}, b_{2\alpha s'})$$
$$\Gamma^1{}_{4\alpha}(b_{1\alpha s}, b_{2\alpha s'})$$

where $s \neq s'$.
The B_2 transformation has the form:

$$B_{2\alpha}(x_1, x_2) = \exp[x_1\Gamma^2{}_{3\alpha}]\exp[x_2\Gamma^2{}_{2\alpha}] \tag{4.62}$$

where

$$\Gamma^2{}_{3\alpha} = \Gamma^1{}_{3\alpha}(b_{1\alpha s}, b_{1\alpha s'})$$
$$\Gamma^2{}_{2\alpha} = \Gamma^1{}_{3\alpha}(b_{2\alpha s}, b_{2\alpha s'})$$
$$\Gamma^2{}_{1\alpha} = \Gamma^1{}_{1\alpha}(b_{2\alpha s}, b_{2\alpha s'})$$
$$\Gamma^2{}_{4\alpha} = \Gamma^1{}_{1\alpha}(b_{1\alpha s}, b_{1\alpha s'})$$
$$\Gamma^2{}_{5\alpha} = \Gamma^1{}_{2\alpha}(b_{2\alpha s}, b_{2\alpha s'})$$
$$\Gamma^2{}_{6\alpha} = \Gamma^1{}_{2\alpha}(b_{1\alpha s}, b_{1\alpha s'})$$

and $s \neq s'$.

4.4.1 Four Space-Time Dimension Case

In four dimensions there are two spin values, up and down. Consequently the combination of spin mixed transformations gives su(2, 2) separately for the b's and the d's for i =1, 2. The su(4, 4) representation has eight real-valued components for the b_i, and $b_i^\dagger$, and d_i, and $d_i^\dagger$, for i = 1, 2 taking account of both spins.[31] Thus the CASe group for four dimension space-time is

$$\mathrm{su}(4, 4) \tag{4.63}$$

The fundamental representation of su(4,4) group has a 16 × 16 matrix form in a 16 real dimension space. Its vectors are analogous to the set of creation/annihilation operators:

$$(b_{1\uparrow}, b_{2\uparrow}, b^\dagger{}_{1\uparrow}, b^\dagger{}_{2\uparrow}, b_{1\downarrow}, b_{2\downarrow}, b^\dagger{}_{1\downarrow}, b^\dagger{}_{2\downarrow}, d_{1\uparrow}, d_{2\uparrow}, d^\dagger{}_{1\uparrow}, d^\dagger{}_{2\uparrow}, d_{1\downarrow}, d_{2\downarrow}, d^\dagger{}_{1\downarrow}, d^\dagger{}_{2\downarrow}) \tag{4.64}$$

Note that the space has four quartets of vectors. This arrangement will be found to persist to the level of the dimension and fermion arrays of NEWQUeST, the octonion

[31] **Note this calculation again requires the use of PseudoQuantum fields.** The four operators (b_i, $b_i^\dagger$, d_i, $d_i^\dagger$) for i = 1,2 and two spin orientations yield 16 real-valued "operator" coordinates. . The fundamental su(4, 4) representation has 16 real-valued coordinates. Thus a match.

If the PseudoQuantum formalism were not used then quantities such as the number of creation/annihilation operators would generally be cut in half. Then the UST would have to be modified by cutting the number of dimensions by a factor of 2 with the result that the number of fundamental fermions would be reduced by a factor of two. The simplest modification of UST in this case would be to remove the Dark fundamental fermions in the UST.

spectrum space of our universe. The subdivisions appear in the known fermion and internal symmetry fundamental representations.

If we consider space-time dimension 4 transformations with spin and b – d mixing then the form of the transformed creation/annihilation (eq. 4.45a) operator becomes

$$b_{1\beta s} = \Sigma_{\alpha,x,s}\,(g_\beta, f_\alpha)\, u^\dagger_{\beta s} u_{\alpha s}\,(c_{11s} b_{1\alpha s} + c_{12s} b_{2\alpha s} + C_{11s} b^\dagger_{1\alpha s} + C_{12} b^\dagger_{2\alpha s} + $$
$$+ c'_{11s} d_{1\alpha s} + c'_{12s} d_{2\alpha s} + C'_{11s} d^\dagger_{1\alpha} + C'_{12s} d^\dagger_{2\alpha})$$

with similar forms for the other transformed operators.

4.4.2 Higher Space-Time Dimension Case

In a space-time of dimension r there are $2^{r/2-1}$ spin values. As a result there is an $su(2^{r/2}, 2^{r/2})$ CASe group containing the direct product subgroup:

$$\bigotimes_{s=1}^{2^{r/2-1}} su(2,2)_s \tag{4.65}$$

where s labels the spin value. Each spin value has an associated su(2, 2) subgroup. For example, for spin ½ there are two subgroups (see above) that combine in an su(4, 4) CASe group.

The combined mixing yields the CASe group for a space-time of r dimensions:

$$su(2^{r/2}, 2^{r/2}) \tag{4.66}$$

with $2^{r/2+2}$ real components for the b_i, $b_i^\dagger$, d_i, and $d_i^\dagger$ for PseudoQuantum fields labeled i = 1, 2 and taking account of all spins.

It is understood that the $su(2^{r/2}, 2^{r/2})$ groups are factors in an infinite tensor product over all momenta.

The number of b's and d's for each momentum, taking account of the PseudoQuantum formalism and spins, is

$$\text{Number of d's and b's} = 2^{r+4}$$

4.4.3 Size of the Fundamental Representations of $su(2^{r/2}, 2^{r/2})$

The matrices of the fundamental representations of $su(2^{r/2}, 2^{r/2})$ are square with $2^{r/2+2}$ rows and $2^{r/2+2}$ columns of real-valued elements. The number of real-valued elements in a matrix is 2^{r+4}. If we use the relation between the space-time dimensions r and the Blaha number N:

$$N = Os = ½(18 - r) \tag{2.30}$$

then we find the number of real-valued elements for an su($2^{r/2}$, $2^{r/2}$) fundamental representation matrix equals the size of the corresponding dimension array d_{dN} in Fig. 1.1:

$$d_{dN} = 2^{22-2N} = 2^{r+4} \tag{4.67}$$
$$= \text{Number of b's and d's for each momentum}$$

Eq. 4.67 is of central importance in the derivation of the HyperCosmos spectrum.

4.5 The HyperCosmos Spectrum of Spaces *Derived*

The derivation of the HyperCosmos spectrum based on the above discussion begins by noting that the space-time dimensions r must be even numbers or eq. 4.66 and the number of b's and d's would both be fractional, and thus inconsistent with the form of second quantized fermion wave functions seen earlier.

Secondly, the lowest *physical* space-time dimension is r = 0. Thus the physical spectrum of HyperCosmos spaces consists of spaces with an even number of space-time dimensions starting at r = 0. The form of the expression for the number of b's and d's in eq. 4.67 limits N to $N \leq 9$ since we assume $r \geq 0$.

The lower bound on $N \geq 0$ follows from the existence of a lower bound on positive energies in the Hydrogen-like atom of the ProtoCosmos.

Thus we obtain the HyperCosmos spectrum of spaces in Figs. 1.1 and 4.1 with:

1. A lower bound: $N \geq 0$
2. A series of spaces with space-time dimensions decreasing by two space by space.
3. An upper bound: $N \leq 9$.

Note that for N = 7 we find r = 4 space-time dimensions and 256 dimensions in the dimension space d_{dN} as seen in QUeST and the Unified SuperStandard Theory (UST)—our universe!

Thus our creation/annihilation operator derivation yields the HyperCosmos spectrum of Figs. 1.1 and 4.1.

We note the choice of parameter in the derivation in section 2.5.4 of chapter 2 agrees with the creation/annihilation operator derivation. We may then view the creation/annihilation operator derivation as determining the choice of parameters.

4.6 The Fermion CASe Groups

The CASe groups of fermions of the universes of The HyperCosmos are determined above in eq. 4.66. See Fig. 4.1.

A General Relativistic transformation to a non-static coordinate system with a different, or without, a Killing vector causes fermion wave function creation/annihilation operators to be CASe transformed in every HyperCosmos space (universe). The correspondence between General Relativistic transformations and the CASe transformations of GiFT reflects the unity of General Relativity and Quantum Theory. See chapter 9.

We take the fermion based derivation of the HyperCosmos spectrum from creation/annihilation operator enumeration as a decisive corroboration of the HyperCosmos derivation in chapters 2 and 3 from an Hydrogen-like atom.

4.7 The N > 9 Spectrum Spaces of Negative Space-time Dimension

The spectrum in Fig. 4.1 contains spaces with negative space-time dimensions. These spaces have fractional totals for dimensions d_{dN}. We will discuss these spaces in chapter 12 where we show negative space-time dimensions have a reasonable mathematical/physical justification. Fractionated particles and dimensions can also be understood physically.

4.8 Anti-Spaces and Anti-Universes

Section 2.10 discusses anti-spaces and anti-universes. We will discuss this topic in more detail in chapter 14.

CASe Groups of the HyperCosmos Spaces (Universes)

	Blaha Space Number $N = o_s$	Cayley-Dickson Number n	Cayley Number d_c	Dimension Array column length d_{dN}	Space-time-Dimension r	CASe Group $su(2^{r/2},2^{r/2})$ CASe
	0	10	1024	2048^2	18	su(512,512)
	1	9	512	1024^2	16	su(256,256)
	2	8	256	512^2	14	su(128,128)
	3	7	128	256^2	12	su(64,64)
	4	6	64	128^2	10	su(32,32)
	5	5	32	64^2	8	su(16,16)
	6	4	16	32^2	6	su(8,8)
	7	**3**	**8**	$\mathbf{16^2}$	**4**	**su(4,4)**
	8	2	4	8^2	2	su(2,2)
	9	1	2	4^2	0	su(1,1)
EXTENSION:						
	10	0	1	2^2	-2	U(1)
	11	-1	½	1^2	-4	U(½)
	12	-2	¼	$½^2$	-6	•
	13	-3	1/8	$¼^2$	-8	•
	14	-4	1/16	$1/8^2$	-10	•

•
•
•

Figure 4.1. The CASe groups of the HyperCosmos spaces.

5. The HyperCosmos Spaces

This chapter describes features of the ten HyperCosmos spaces and universes of Fig. 1.2. Nine spaces, N = 1, 2, … , 9, can be viewed as subspaces (Fig. 5.1) of the N = 0 space although we will view them as independent here. See Fig. 5.3. In chapter 8 we will consider the generation of universes by fermion-antifermion annihilation.

5.1 Space N = 0

The N = 0 space/universe is generated in the ProtoCosmos as described in chapter 3. It generates N = 1, 2 , 3, 4, … 9 universes by fermion-antifermion annihilation in the model of chapter 8.

5.2 The Maxiverse N = 5

The Maxiverse is the proposed parent of the Multiverse (Megaverse) of our universe. It has 2^{12} = 4096 total dimensions of which 8 are real-valued space-time dimensions. The Multiverse(s) is generated from it by fermion-antifermion annihilation.

5.3 The Multiverse (Megaverse) N = 6

The Multiverse is the parent of our universe and possibly of other universes. It has 2^{10} = 1024 total dimensions of which 6 are real-valued space-time dimensions. There may be several Multiverses (and associated universes.)

5.4 N = 7 Universes such as Our Universe

Our universe is an N = 7 universe with four real-valued space-time dimensions.

5.5 Spaces N = 8 and 9

These spaces may or may not have associated universes. At present there is no evidence in our universe of space wave function fermions or bosons as described in chapter 6. Consequently the possibility of their creation is doubtful.

5.6 Spaces of N ≥ 10

These spaces have negative space-time dimensions, which we showed to be physically understandable in Blaha (2022b). They are used for fractionation as shown in chapter 12. They originate in the atomic energy eigenstates.

5.7 Anti-Spaces and Universes

The atom of chapter 2 has negative energy eigenstates. These states correspond to anti-spaces and anti-universes in the HyperCosmos space spectrum. The HyperCosmos space and anti-space spectrums are identical. Universes have the spaces spectrum and positive charge. Anti-universes have the same spectrum (Fig. 5.4) but contain anti-particles of negative charge as described in section 2.11 and chapter 6.

5.8 Hierarchies of Universes

It is possible to generate hierarchies of universes through multiple fermion-antifermion annihilations. The simplest hierarchy consists of one universe creation for each of the top eight spaces. It is also possible to create a hierarchy tree as shown in the example of Fig. 5.2. Hierarchies can contain both universes and anti-universes.

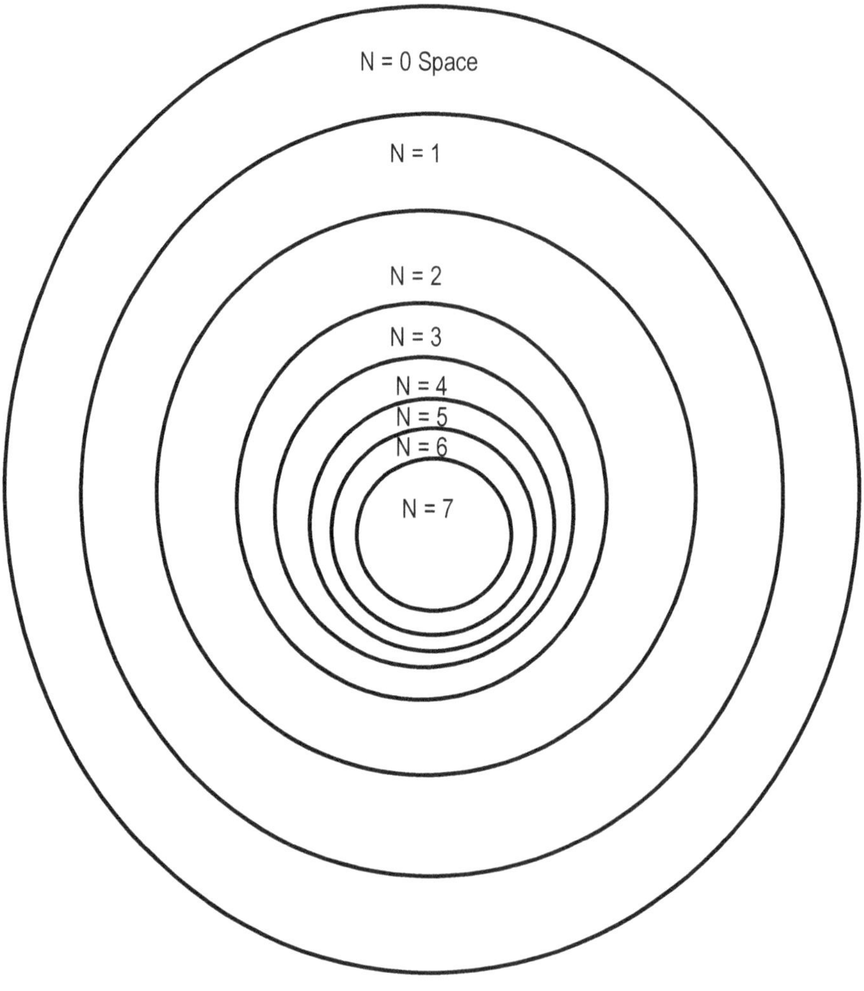

Figure 5.1. The eight spaces most relevant for the HyperCosmos. Our universe is an N = 7 space.

A COSMOS

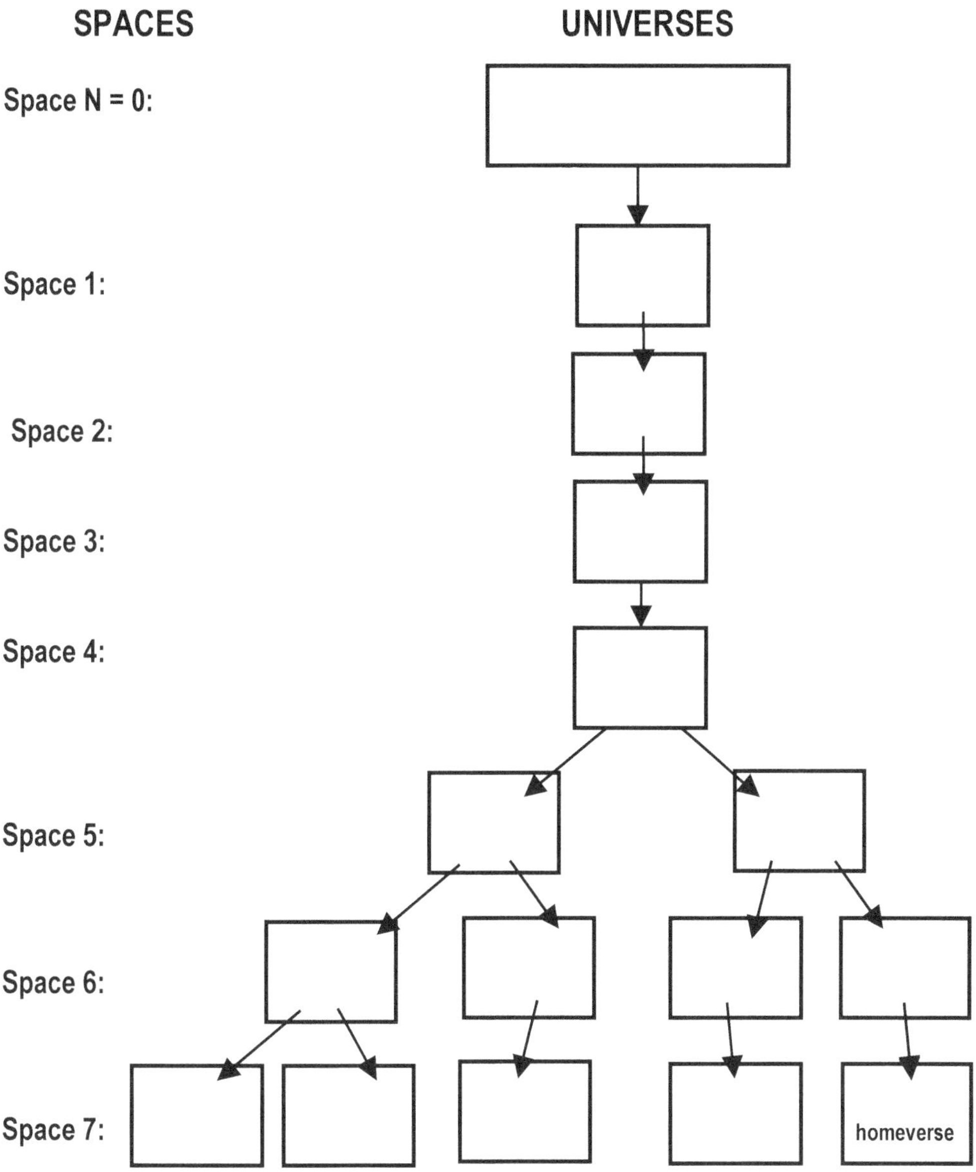

Figure 5.2. A hierarchy of universes (and possibly anti-universes) leading from the N = 0 space to the "homeverse" – our designation for our N = 7 universe. The homeverse has one "sibling" and three "cousin" universes. *The entire hierarchy resides in the N = 0 universe since the inheritance stems from the N = 0 universe. Other universes could be "reached" from the N = 0 universe if a mode of transportation exists.*

THE HYPERCOSMOS SPACES SPECTRUM

Blaha Space Number $N = o_s$	Cayley-Dickson Number n	Cayley Number d_c	Dimension Array column length d_{dN}	Space-time-Dimension r	CASe Group $su(2^{r/2}, 2^{r/2})$ CASe
0	10	1024	2048^2	18	su(512,512)
1	9	512	1024^2	16	su(256,256)
2	8	256	512^2	14	su(128,128)
3	7	128	256^2	12	su(64,64)
4	6	64	128^2	10	su(32,32)
5	5	32	64^2	8	su(16,16)
6	4	16	32^2	6	su(8,8)
7	**3**	**8**	$\mathbf{16^2}$	**4**	**su(4,4)**
8	2	4	8^2	2	su(2,2)
9	1	2	4^2	0	su(1,1)
10	0	1	2^2	-2	U(1)
11	-1	½	1^2	-4	U(½)
12	-2	¼	$½^2$	-6	•
13	-3	1/8	$¼^2$	-8	•
14	-4	1/16	$1/8^2$	-10	•

•
•
•

Figure 5.3. The HyperCsmos spectrum of spaces. The spaces spectrum corresponds to the positive energy solutions of the Hydrogen-like atom.

THE HYPERCOSMOS ANTI-SPACES SPECTRUM

Blaha Space Number $N = o_s$	Cayley-Dickson Number n	Cayley Number d_c	Dimension Array column length d_{dN}	Space-time-Dimension r	CASe Group $su(2^{r/2},2^{r/2})$ CASe
0	10	1024	2048^2	18	su(512,512)
1	9	512	1024^2	16	su(256,256)
2	8	256	512^2	14	su(128,128)
3	7	128	256^2	12	su(64,64)
4	6	64	128^2	10	su(32,32)
5	5	32	64^2	8	su(16,16)
6	4	16	32^2	6	su(8,8)
7	**3**	**8**	$\mathbf{16^2}$	**4**	**su(4,4)**
8	2	4	8^2	2	su(2,2)
9	1	2	4^2	0	su(1,1)
10	0	1	2^2	-2	U(1)
11	-1	½	1^2	-4	U(½)
12	-2	¼	$½^2$	-6	•
13	-3	1/8	$¼^2$	-8	•
14	-4	1/16	$1/8^2$	-10	•

•
•
•

Figure 5.4. The HyperCosmos spectrum for anti-spaces. Spaces and anti-spaces have the same spectrum. The anti-spaces spectrum corresponds to the negative energy solutions of the Hydrogen-like atom.

6. HyperCosmos Space Free Wave Functions for Fermions and Bosons

The spaces of the HyperCosmos require fermion and boson second quantized wave functions for elementary particles. We will use them in chapter 8 in fermion-antifermion annihilation to generate universes. The generation of universes requires a new type of wave function for fermions and bosons. We call these new particle wave functions *space wave functions*. We considered possible forms for them in earlier books. This chapter introduces them through a new mechanism based on the creation and annihilation operators[32] in free field expansions. For example eqs. 6.1 and 6.2 below define these operators jointly for a host space-time and an *inner universe* contained within host space-time states.[33]

The key to space particle wave functions is in the form of the wave function expansion in joint *host particle* – inner universe particle creation and annihilation operators. This chapter and chapter 4 explore this new approach. Chapter 7 examines the possibility of protons with quarks confined as inner "universes".

6.1 HyperCosmos Free Fermion Space Wave Functions

We consider fermions in a Blaha number N_1 HyperCosmos (*host*) universe that contain an N_2 (*inner*) space/universe. We will use the GiFT PseudoQuantum formalism.

6.1.1 Host Space Fermion without Host Internal Symmetries

We begin by defining creation and annihilation operators for this space (universes) configuration for a fermion without internal symmetries in the N_1 universe. Later we consider a fermion with internal symmetries of the N_1 space.

$$b_{iN_1N_2\alpha M}(k, q, s)^\dagger \quad \text{and} \quad b_{i\,N_1N_2\alpha M}(k, q, s) \tag{6.1}$$

for PseudoQuantum i = 1, 2, where s is the spin, M is the mass-energy of the created universe, N_1 is the host Blaha number, $r1 = 18 - 2N_1$ (by eq. 2.30)) is the number of space-time dimensions of the *host* universe with host momentum $k = (k_1, k_2, \ldots, k_{r1})$. We set the mass of the host fermion to be m. The coordinates of the host space are $x = (x_1, x_2, \ldots, x_{r1})$ with $t = x_{r1}$. The coordinates of the inner space (universe) are $u = (u_1, u_2, \ldots, u_{r2})$. The inner universe momentum is $q = (q_1, q_2, \ldots, q_{r2})$.

The space-time dimensions of the host space and the inner universe are different in general. Either may have a greater space-time dimension.[34]

[3232] The study of these operators in this book and earlier books indicates that much remains to be learned from their deeper consideration. Our work is a beginning.

[33] An alternative considered in earlier books does not seem as good.

[34] We use r1 and r2 for space-time dimensions, rather than with subscripts, for typographic convenience to avoid miniscule subscripting.

Within the fermion we have an inner space (universe) of dimension $d_{dN_2} = 2^{22 - 2N_2}$, which contains r2 = 18 – 2N₂ real-valued space-time. The index α labels the dimensions of the inner N_2 space. We take $t_u = u_{r2}$ to be the time variable within the inner universe. The index α labels non-space-time components for the d_{dN2} – r2 real-valued dimension internal symmetry group representation for the inner universe.

The momentum k is for the host fermion. The momentum q is an r2-vector for the inner space (universe). We similarly define hole creation and annihilation operators for anti-spaces (anti-universes).

$$d_{iN_1N_2\alpha M}(k, q, s)^\dagger \quad \text{and} \quad d_{iN_1N_2\alpha M}(k, q, s) \tag{6.2}$$

The non-zero anticommutator relations are

$$\{b_{iN_1N_2\alpha M}(k, q, s), b_{jN_1N_2\alpha' M}(k', q', s')^\dagger\} = \delta_{ss'}\delta_{\alpha\alpha'}\delta(\mathbf{k} - \mathbf{k'})\, \delta^{r1-1}(\mathbf{q} - \mathbf{q'}) \tag{6.3}$$
$$\{d_{iN_1N_2\alpha M}(k, q, s), d_{jN_1N_2\alpha' M}(k', q', s')^\dagger\} = \delta_{ss'}\delta_{\alpha\alpha'}\delta(\mathbf{k} - \mathbf{k'})\, \delta^{r1-1}(\mathbf{q} - \mathbf{q'}) \tag{6.4}$$

for i, j = 1, 2 with i ≠ j.

The free fermion PseudoQuantum wave functions are:

$$\psi_{iN_1N_2\alpha M}(\mathbf{x}, t, \mathbf{u}, t_u) = \Sigma_s \int d^{r1-1}k \int d^{r2-1}q\, \mathfrak{N}(q, k)[u_{r1}(k, s)\, b_{iN_1N_2\alpha M}(k, q, s)e^{-i(k\cdot x + q\cdot u)} + $$
$$+ v_{r1}(k, s)d_{iN_1N_2\alpha M}(k, q, s)^\dagger e^{i(k\cdot x + +q\cdot u)}] \tag{6.5}$$

for i = 1, 2 where $\mathfrak{N}$ is a normalization constant, where m is the host fermion mass in the N_1 space's space-time and M is the mass-energy of the universe space of Blaha number N_2. We assume the $k^2 = m^2$ of the host fermion and $q^2 = M^2$ in the inner $N_2 = 0$ space. Note $x = (x^0, \mathbf{x})$, $u = (u^0, \mathbf{u})$, $t = x^0$, and $t_u = u^0$. $\psi_{iN_1N_2\alpha M}(x, u)$ forms a representation of su(2^{9-N}, 2^{9-N}) (by chapter 4) if one includes *both* internal symmetries and the inner universe's space-time. See Figs. 1.1 and 1.2.

We tentatively define the Lagrangian for the host fermion space particle interacting with a boson space particle as:

$$\mathcal{L} = \overline{\psi}(x, u)[i\gamma^\mu\partial/\partial x^\mu - g'\varphi - m + (\partial/\partial u^\mu\, \partial/\partial u_\mu - M^2)/\lambda]\psi(x, u) +$$

$$+ \tfrac{1}{2}(\partial/\partial x^\mu\varphi(x, u)\, \partial/\partial x_\mu\varphi(x, u) - m'^2\varphi^2(x, u) + (\partial/\partial u^\mu\, \varphi(x, u)\partial/\partial u_\mu\, \varphi(x, u) - M'^2\varphi^2(x, u))/\lambda) \tag{6.6}$$

where m′ is the mass of the host φ space boson, M′ is the mass of the inner φ space boson, and λ is a constant mass[35]. The fermion dynamic equation is

$$\delta\mathcal{L}/\delta\overline{\psi} = [i\gamma^\mu\partial/\partial x^\mu - g'\varphi - m + (\partial/\partial u^\mu\, \partial/\partial u_\mu - M^2)/\lambda]\, \psi(x, u) = 0 \tag{6.6a}$$

[35] It is irrelevant after the separation of equations.

We separate this fermion equation to obtain a host fermion equation and a inner universe equation.[36] The *host*, free field, fermion Dirac equation is[37]

$$(i\gamma^{\mu}\partial/\partial x^{\mu} - m)\psi(x, u) = 0 \tag{6.7}$$

The inner space/universe of the fermion has a Klein-Gordan equation under the assumption that it is a particle:[38] Its equation for the mass-energy M is

$$(\partial/\partial u^{\mu}\ \partial/\partial u_{\mu} - M)\psi(x, u) = 0 \tag{6.8}$$

The two differential equations hold separately for the host and inner space coordinates.

The host fermion spinors u_{r1} and v_{r2} satisfy

$$\Sigma_s\ u_{r1}(k, s)_{\sigma}\bar{u}_{r1}(k, s)_{\tau} = (\not{k} + m)_{\sigma\tau}/(2m) \tag{6.9}$$

$$\Sigma_s\ v_{r1}(k, s)_{\sigma}\bar{v}_{r1}(k, s)_{\tau} = (m - \not{k})_{\sigma\tau}/(2m) \tag{6.10}$$

for spinor indices σ and τ for spinors in any host space-time dimension.[39] The number of components of a spinor column is 2^{9-N} and the host fermion spin is

$$s = (2^{8-N} - 1)/2 \tag{6.11}$$

For N = 7 (our universe) the number of components of a column is 4 and s = ½.

The wave function equal time anti-commutator is

$$\{\psi_{iN_1N_2\alpha M}(\mathbf{x}, t, \mathbf{u}, t_u), \psi_{jN_1N_2\alpha' M}(\mathbf{x}', t, \mathbf{u}', t_u)^{\dagger}\gamma^0\} = \int d^{r1-1}k \int d^{r2-1}q \int d^{r1-1}k' \int d^{r2-1}q'\ \mathfrak{N}(q, k)\ \mathfrak{N}(q', k')\cdot$$

$$\cdot\{[b_{iN_1N_2\alpha M}(k, q)e^{-i(k\cdot x+q\cdot u)} + d_{iN_1N_2\alpha M}{}^{\dagger}(k, q)e^{i(k\cdot x+q\cdot u)}], [b_{jN_1N_2\alpha' M}(k', q')^{\dagger}e^{i(k'\cdot x'+q'\cdot u')} + d_{jN_1N_2\alpha' M}(k', q')e^{-i(k'\cdot x'+q'\cdot u')}]\ \gamma^0\}$$

$$= \Sigma_{s,s'} \int d^{r1-1}k \int d^{r2-1}q \int d^{r1-1}k' \int d^{r2-1}q'\ \mathfrak{N}(q, k)\ \mathfrak{N}(q', k')\delta_{\alpha\alpha'}\delta^{r1-1}(k - k')\ \delta^{r2-1}(q - q')\cdot$$

$$\cdot[e^{-i(k\cdot x+q\cdot u)}e^{i(k'\cdot x' + q'\cdot u')}\ u_{r1}(k, s)\overline{u_{r1}}(k', s') + e^{i(k\cdot x+q\cdot u)}e^{-i(k'\cdot x' + q'\cdot u')}\ v_{r1}(k, s)_{\sigma}\overline{v_{r1}}(k', s')]$$

$$= \int d^{r1-1}k \int d^{r2-1}q \int d^{r1-1}k' \int d^{r2-1}q'\ \mathfrak{N}(q, k)\ \mathfrak{N}(q', k')\delta_{\alpha\alpha'}\delta^{r1-1}(k - k')\ \delta^{r2-1}(q - q')\cdot$$

$$\cdot[e^{-i(k\cdot x+q\cdot u)}e^{i(k'\cdot x' + q'\cdot u')}\ (\not{k} + m)/(2m) - e^{i(k\cdot x+q\cdot u)}e^{-i(k'\cdot x' + q'\cdot u')}\ (m - \not{k})/(2m)]$$

$$= \delta_{\alpha\alpha'}\int d^{r1-1}k \int d^{r2-1}q\ (2k^0/m)\ \mathfrak{N}(q, k)^2\ [e^{-i(k\cdot x+q\cdot u)}e^{i(k\cdot x' + q\cdot u')} + e^{i(k\cdot x+q\cdot u)}e^{-i(k\cdot x' + q\cdot u')}]$$

---

[36] The classical ψ(x, u) is separable. The quantum ψ(x, u) is not separable because of the form of the creation and annihilation operators $b_{iN_1N_2\alpha M}(k, q, s)^{\dagger o}$, which are so defined to unite the host fermion with the embedded universe.

[37] Thie basis of the higher dimension fermions is described in detail in Blaha (2021g).

[38] See Blaha (2021d) *Universes are Particles.*

[39] Blaha (2021d).

for $i \neq j$ where the space's time is t, the independent embedded universe's time is t_u, and γ^0 is a Dirac matrix for the host space-time. Note the spatial parts are in bold typeface. Note also $t \neq t_u$. They are independent variables. If

$$\mathfrak{N}(q, k) = (2k^0/m)^{-1/2}(2\pi)^{-(r1-1)/2-(r2-1)/2}$$

then

$$\{\psi_{iN_1N_2\alpha M}(\mathbf{x}, t, \mathbf{u}, t_u), \psi_{jN_1N_2\alpha' M}(\mathbf{x}', t, \mathbf{u}', t_u)^\dagger\gamma^0\} = \delta_{\alpha\alpha'}\delta^{r1-1}(\mathbf{x} - \mathbf{x}')\,\delta^{r2-1}(\mathbf{u} - \mathbf{u}') \quad (6.12)$$

with other anti-commutators zero.

The Feynman propagator (with $t = u^0$ and $t' = u'^0$) is:

$$\begin{aligned}
S_F(x' - x, u' - u)\,\gamma^0 &= -i\langle 0|\,\psi_{iN_1N_2\alpha M}(\mathbf{x}', t', \mathbf{u}', t_{u'})\gamma^0\psi_{iN_1N_2\alpha M}(\mathbf{x}, t, \mathbf{u}, t_u)^\dagger|0\rangle\theta(t' - t) + \\
&\quad + i\,\langle 0|\,\psi_{iN_1N_2\alpha M}(\mathbf{x}, t, \mathbf{u}, t_u)^\dagger\gamma^0\psi_{iN_1N_2\alpha M}(\mathbf{x}', t', \mathbf{u}', t_{u'})|0\rangle\theta(t - t') \\
&= \Sigma_{s,s'}\int d^{r1-1}k\int d^{r2-1}q\int d^{r1-1}k'\int d^{r2-1}q'\;\mathfrak{N}(q, k)\,\mathfrak{N}(q', k')\delta_{\alpha\alpha'}\,\delta^{r1-1}(k - k')\,\delta^{r2-1}(q - q')\cdot \\
&\quad \cdot[e^{i(k\cdot x+q\cdot u)}e^{-i(k'\cdot x' + q'\cdot u')}\,u_{r1}(k', s')\overline{u}_{r1}(k, s)\theta(t' - t) + \\
&\quad + e^{-i(k\cdot x+q\cdot u)}e^{i(k'\cdot x' + q'\cdot u')}\,\overline{v}_{r1}(k, s)v_{r1}(k', s')\,\theta(t - t')] \\
&= \Sigma_s\int d^{r1-1}k\int d^{r2-1}q\;\mathfrak{N}(q, k)^2\delta_{\alpha\alpha'}[e^{i(k\cdot(x - x') + q\cdot(u - u'))}u_{r1}(k, s')\overline{u}_{r1}(k, s)\theta(t' - t) + \\
&\quad + e^{i(k\cdot(x' - x) + q\cdot(u' - u))}\,v_{r1}(k, s)\overline{v}_{r1}(k, s)\,\theta(t - t')] \\
&= \int d^{r1-1}k\int d^{r2-1}q\;\mathfrak{N}(q, k)^2\delta_{\alpha\alpha'}[e^{i(k\cdot(x - x') + q\cdot(u - u'))}\,u_{r1}(k, s')u_{r1}(\overline{k}, s)\theta(t' - t) + \\
&\quad + e^{i(k\cdot(x' - x) + q\cdot(u' - u))}\,v_{r1}(k, s)\overline{v}_{r1}(k, s')\,\theta(t - t')] \\
&= \delta_{\alpha\alpha'}\int d^{r1-1}k\int d^{r2-1}q\,(4k^0)^{-1}(2\pi)^{-(r1-1)-(r2-1)}\,[e^{i(k\cdot(x - x') + q\cdot(u - u'))}\,(\not{k} + m)\theta(t' - t) + \\
&\quad + e^{-i(k\cdot(x' - x) + q\cdot(u' - u))}\,(m - \not{k})\theta(t - t')]
\end{aligned} \quad (6.13)$$

If $u^0 = u^{0\prime}$ the universe part factors out, the inner universe is unchanged, and the host part is Feynman-like:

$$\begin{aligned}
S_F(x' - x)\,\gamma^0 &= \delta^{r2-1}(\mathbf{u} - \mathbf{u}')\,\delta_{\alpha\alpha'}\int d^{r1-1}k\,(4k^0)^{-1}(2\pi)^{-(r1-1)}\,[e^{ik\cdot(x - x')}(\not{k} + m)\theta(t' - t) + \\
&\quad + e^{ik\cdot(x' - x)}(m - \not{k})\theta(t - t')] \\
&= \delta^{r2-1}(\mathbf{u} - \mathbf{u}')\,\delta_{\alpha\alpha'}\int d^{r1}k(2\pi)^{-r1}\,e^{-ik\cdot(x' - x)}/(\not{k} - m + i\varepsilon)
\end{aligned} \quad (6.14)$$

If $u^0 \neq u^{0\prime}$ then from eq. 6.13 we have

$$S_F(x' - x, u' - u)\gamma^0 = \delta_{\alpha\alpha'} \int d^{r1-1}k \int d^{r2-1}q\, (4k^0)^{-1}(2\pi)^{-(r1-1)-(r2-1)}[e^{i(k\cdot(x-x')+q\cdot(u-u'))}(\not{k}+m)\theta(t'-t) +$$

$$+ e^{-i(k\cdot(x'-x)+q\cdot(u'-u))}(m - \not{k})\theta(t-t')] \qquad (6.15)$$

For $u^0 \neq u^{0\prime}$ we can express S_F in terms of the homogeneous wave equation solutions in r space-time dimensions:

$$\Delta_{rm}(y' - y) = -i \int d^{r-1}k\, (e^{-ik\cdot(y'-y)} - e^{ik\cdot(y'-y)})/[(2\pi)^{r-1}2\omega_k] \qquad (6.16)$$

$$\Delta_{1rm}(y' - y) = \int d^{r-1}k\, (e^{-ik\cdot(y'-y)} + e^{ik\cdot(y'-y)})/[(2\pi)^{r-1}2\omega_k] \qquad (6.17)$$

$$\Delta_{2rm}(y' - y) = \int d^{r-1}k\, e^{-ik\cdot(y'-y)}/[(2\pi)^{r-1}2\omega_k] = ½(\Delta_1(y' - y) + i\,\Delta(y' - y)) \qquad (6.18)$$

$$\Delta_{3rm}(y' - y) = \int d^{r-1}k\, e^{+ik\cdot(y'-y)}/[(2\pi)^{r-1}2\omega_k] = ½(\Delta_1(y' - y) - i\,\Delta(y' - y)) \qquad (6.19)$$

where $\omega_k = (\mathbf{k}^2 + m^2)^{½}$. Then

$$S_F(x' - x, u' - u)\gamma^0 = \delta_{\alpha\alpha'}\{(-i\gamma\cdot\partial + m)\Delta_{3r1m}(x' - x)\partial/\partial u^0\Delta_{3r2M}(u' - u)\theta(t' - t) +$$

$$+ (-i\gamma\cdot\partial - m)\Delta_{2r1m}(x' - x)\partial/\partial u^0\Delta_{2r2M}(u' - u)\theta(t - t')\} \qquad (6.20)$$

The host fermion and the inner universe propagate "jointly" with an interaction between the Host propagation and the inner universe propagation.

A one host-universe state is created from the vacuum by

$$|N_1N_2\alpha Mkqs> = b_{2N_1N_2\alpha M}(k, q, s)^\dagger|0>_2 \qquad (6.21)$$

using the vacuum $|0>_2$ of PseudoQuantum theory. The host-universe initial wave function is

$$\psi_{N_1N_2\alpha M}(x, u, k, q, s) = <x, u|\, N_1N_2\alpha Mkqs> \qquad (6.22)$$

Note that a superposition of wave functions using a distribution of momenta, and mass M introduces a further quantum aspect in the initial universe particle state.

Interaction terms such as the electromagnetic-like interaction for atoms given earlier can be defined.

6.1.2 Host Space Fermion with Host Internal Symmetries

This section is similar to section 6.1.1. It introduces internal symmetry in the *host* space-time. The symmetry will correspond to the Blaha number N_1 space with $d_{dN_1} = 2^{22-2N_1}$ total dimensions by eq. 2.25. We introduce a corresponding index β that ranges from 1 to $2^{22-2N_1} - r1$ where $r1 = 18 - 2N_1$ is the number of space-time dimensions.

The space and space-time dimensions of the host space and the inner universe are different in general. Either may have a greater space-time dimension.

We now repeat the discussion by displaying equations of section 6.1.1 modified to have a β index. Typically α → αβ.

$$b_{iN_1N_2\alpha\beta M}(k, q, s)^\dagger \quad \text{and} \quad b_{i\,N_1N_2\alpha\beta M}(k, q, s) \qquad (6.23)$$

for PseudoQuantum i = 1, 2, where s is the spin, M is the mass-energy of the created universe, N_1 is the host Blaha number, $r1 = 18 - 2N_1$ (by eq. 2.30)) is the number of space-time dimensions of the *host* universe with momentum $k = (k_1, k_2, \ldots, k_{r1})$, and with $t = k_{r1}$. We set the mass of the host fermion to be m. The coordinates of the host space are $x = (x_1, x_2, \ldots, x_{r1})$.

The space-time dimensions of the host space and the inner universe are different in general. Either may have a greater space-time dimension.

Within the fermion we have an inner space of dimension $d_{dN_2} = 2^{22 - 2N_2}$, which contains an $r2 = 18 - 2N_2$ real-valued space-time. The index α labels the dimensions of the inner N_2 space. We take $t_u = u_{r1}$ to be the time variable within the inner universe. The index α labels non-spacetime components for the $d_{dN2} - r2$ real-valued dimension internal symmetry group representation for the inner universe. The coordinates of the host space are $u = (u_1, u_2, \ldots, u_{r2})$ with $t_u = u_{r2}$. The *host* universe momentum is $q = (q_1, q_2, \ldots, q_{r2})$.

The momentum k is for the host fermion. The momentum q is an r2-vector for the embedded space (universe). We similarly define hole creation and annihilation operators for anti-spaces (anti-universes).

$$d_{iN_1N_2\alpha\beta M}(k, q, s)^\dagger \quad \text{and} \quad d_{iN_1N_2\alpha\beta M}(k, q, s) \qquad (6.24)$$

The non-zero anticommutator relations are

$$\{b_{iN_1N_2\alpha\beta M}(k, q, s), b_{jN_1N_2\alpha'\beta'M}(k', q', s')^\dagger\} = \delta_{ss'}\delta_{\alpha\alpha'}\delta_{\beta'\beta}\delta(\mathbf{k} - \mathbf{k}')\,\delta^{r1-1}(\mathbf{q} - \mathbf{q}') \qquad (6.25)$$
$$\{d_{iN_1N_2\alpha\beta M}(k, q, s), d_{jN_1N_2\alpha'\beta'M}(k', q', s')^\dagger\} = \delta_{ss'}\delta_{\alpha\alpha'}\delta_{\beta'\beta}\delta(\mathbf{k} - \mathbf{k}')\,\delta^{r1-1}(\mathbf{q} - \mathbf{q}') \qquad (6.26)$$

for i, j = 1, 2 with i ≠ j.

The free PseudoQuantum wave functions are:

$$\psi_{iN_1N_2\alpha\beta M}(\mathbf{x}, t, \mathbf{u}, t_u) = \Sigma_s \int d^{r1-1}k \int d^{r2-1}q\, \mathcal{N}(q, k)[u_{r1}(k, s)\, b_{iN_1N_2\alpha\beta M}(k, q, s)e^{-i(k\cdot x + q\cdot u)} + $$
$$+ v_{r1}(k, s)d_{iN_1N_2\alpha\beta M}(k, q, s)^\dagger e^{i(k\cdot x + +q\cdot u)}] \qquad (6.27)$$

for i = 1, 2 where $\mathcal{N}$ is a normalization constant, where m is the host fermion mass in the N_1 space's space-time and M is the mass-energy of the universe space of Blaha number N_2. We assume the $k^2 = m^2$ of the host fermion and $q^2 = M^2$ in the inner $N_2 = 0$ space. Note $x = (x^0, \mathbf{x})$, $u = (u^0, \mathbf{u})$, $t = x^0$, and $t_u = u^0$. $\psi_{iN_1N_2\alpha\beta M}(x, u, k , q, s)$ forms a representation of su(2^{9-N}, 2^{9-N}) (by chapter 4) if one includes *both* internal symmetries and the inner universe's space-time. See Figs. 1.1 and 1.2.

We define the Lagrangian for the host fermion space particle interacting with a boson space particle as:

$$\mathcal{L} = \bar{\psi}(x, u)[i\gamma^\mu\partial/\partial x^\mu - g'\varphi - m + (\partial/\partial u^\mu\, \partial/\partial u_\mu - M^2)/\lambda]\psi(x, u) +$$

$$+ \tfrac{1}{2}(\partial/\partial x^\mu\varphi(x, u)\, \partial/\partial x_\mu\varphi(x, u) \;- m'^2\varphi^2(x, u) + (\partial/\partial u^\mu\, \varphi(x, u)\partial/\partial u_\mu\, \varphi(x, u) - M'^2\varphi^2(x, u))/\lambda) \tag{6.28}$$

where m′ is the mass of the host φ space boson, M′ is the mass of the inner φ space boson, and λ is a constant mass[40]. The fermion dynamic equation is

$$\delta\mathcal{L}/\delta\bar{\psi} = [i\gamma^\mu\partial/\partial x^\mu - g'\varphi - m + (\partial/\partial u^\mu\, \partial/\partial u_\mu - M^2)/\lambda]\, \psi(x, u) = 0 \tag{6.29}$$

We separate this fermion equation to obtain a host fermion equation and an inner universe equation.[41] The *host*, free field, fermion Dirac equation is[42]

$$(i\gamma^\mu\partial/\partial x^\mu - m)\psi(x, u) = 0 \tag{6.30}$$

The inner space/universe of the fermion has a Klein-Gordan equation under the assumption that it is a particle:[43] Its equation for the mass-energy M is

$$(\partial/\partial u^\mu\, \partial/\partial u_\mu - M^2)\psi(x, u) = 0 \tag{6.31}$$

The two differential equations hold separately for the host and inner space coordinates.

The host fermion spinors u_{r1} and v_{r2} satisfy

$$\Sigma_s\, u_{r1}(k, s)_\sigma \bar{u}_{r1}(k, s)_\tau = (\not{k} + m)_{\sigma\tau}/(2m) \tag{6.32}$$

$$\Sigma_s\, v_{r1}(k, s)_\sigma \bar{v}_{r1}(k, s)_\tau = (m - \not{k})_{\sigma\tau}/(2m) \tag{6.33}$$

for spinor indices σ and τ for spinors in any space-time dimension.[44] The number of components of a spinor column is 2^{9-N} and the host fermion spin is

$$s = (2^{8-N} - 1)/2 \tag{6.34}$$

For N = 7 (our universe) the number of components of a column is 4 and s = ½.

---

40 It is irrelevant after the separation of equations.

41 The classical ψ(x, u) is separable. The quantum ψ(x, u) is not separable because of the form of the creation and annihilation operators $b_{iN1N2\alpha M}(k, q, s)^{\dagger o}$, which are so defined to unite the host fermion with the embedded universe.

42 Thie basis of the higher dimension fermions is described in detail in Blaha (2021g).

43 See Blaha (2021d) *Universes are Particles.*

44 Blaha (2021d).

The wave function equal time anti-commutator is

$$\{\psi_{iN_1N_2\alpha\beta M}(\mathbf{x}, t, \mathbf{u}, t_u), \psi_{jN_1N_2\alpha'\beta'M}(\mathbf{x}', t, \mathbf{u}', t_u)^\dagger\gamma^0\} = \int d^{r1-1}k \int d^{r2-1}q \int d^{r1-1}k' \int d^{r2-1}q'\, \mathfrak{N}(q, k)\, \mathfrak{N}(q', k')\cdot$$

$$\cdot\{[b_{iN_1N_2\alpha\beta M}(k, q)e^{-i(k\cdot x+q\cdot u)} + d_{iN_1N_2\alpha\beta M}{}^\dagger(k, q)e^{i(k\cdot x+q\cdot u)}], [b_{jN_1N_2\alpha'\beta'M}(k', q')^\dagger e^{i(k'\cdot x'+q'\cdot u')} + d_{jN_1N_2\alpha'\beta'M}(k', q')\cdot$$
$$\cdot e^{-i(k'\cdot x'+q'\cdot u')}]\,\gamma^0\}$$

$$= \Sigma_{s,s'} \int d^{r1-1}k \int d^{r2-1}q \int d^{r1-1}k' \int d^{r2-1}q'\, \mathfrak{N}(q, k)\, \mathfrak{N}(q', k')\delta_{\alpha\alpha'}\delta_{\beta\beta'}\delta^{r1-1}(k - k')\, \delta^{r2-1}(q - q')\cdot$$

$$\cdot[e^{-i(k\cdot x+q\cdot u)}e^{i(k'\cdot x' + q'\cdot u')}\, u_{r1}(k, s)\overline{u}_{r1}(k', s') + e^{i(k\cdot x+q\cdot u)}e^{-i(k'\cdot x' + q'\cdot u')}\, v_{r1}(k, s)_\sigma\overline{v}_{r1}(k', s')]$$

$$= \int d^{r1-1}k \int d^{r2-1}q \int d^{r1-1}k' \int d^{r2-1}q'\, \mathfrak{N}(q, k)\, \mathfrak{N}(q', k')\delta_{\alpha\alpha'}\delta_{\beta\beta'}\delta^{r1-1}(k - k')\, \delta^{r2-1}(q - q')\cdot$$

$$\cdot[e^{-i(k\cdot x+q\cdot u)}e^{i(k'\cdot x' + q'\cdot u')}\, (\not{k} + m)/(2m) - e^{i(k\cdot x+q\cdot u)}e^{-i(k'\cdot x' + q'\cdot u')}\, (m - \not{k})/(2m)]$$

$$= \delta_{\alpha\alpha'}\delta_{\beta\beta'}\int d^{r1-1}k \int d^{r2-1}q\, (2k^0/m)\, \mathfrak{N}(q, k)^2\, [e^{-i(k\cdot x+q\cdot u)}e^{i(k\cdot x' + q\cdot u')} + e^{i(k\cdot x+q\cdot u)}e^{-i(k\cdot x' + q\cdot u')}]$$

for $i \neq j$ where the space's time is t, the independent embedded universe's time is t_u, and γ^0 is a Dirac matrix for the host space-time. Note the spatial parts are in bold typeface. Note also $t \neq t_u$. They are independent variables. If

$$\mathfrak{N}(q, k) = (2k^0/m)^{-\frac{1}{2}}(2\pi)^{-(r1 - 1)/2 - (r2 - 1)/2}$$

then

$$\{\psi_{iN_1N_2\alpha\beta M}(\mathbf{x}, t, \mathbf{u}, t_u), \psi_{jN_1N_2\alpha'\beta'M}(\mathbf{x}', t, \mathbf{u}', t_u)^\dagger\gamma^0\} = \delta_{\alpha\alpha'}\, \delta_{\beta\beta'}\delta^{r1-1}(\mathbf{x} - \mathbf{x}')\, \delta^{r2-1}(\mathbf{u} - \mathbf{u}') \quad (6.35)$$

with other anti-commutators zero.

The Feynman propagator (with $t = u^0$ and $t' = u'^0$) is:

$$S_F(x' - x, u' - u)\,\gamma^0 = -i\langle 0|\, \psi_{iN_1N_2\alpha'\beta'M}(\mathbf{x}', t', \mathbf{u}', t_{u'})\gamma^0\psi_{iN_1N_2\alpha\beta M}(\mathbf{x}, t, \mathbf{u}, t_u)^\dagger|0\rangle\theta(t' - t) +$$
$$+ i\langle 0|\, \psi_{iN_1N_2\alpha\beta M}(\mathbf{x}, t, \mathbf{u}, t_u)^\dagger\gamma^0\psi_{iN_1N_2\alpha'\beta'M}(\mathbf{x}', t', \mathbf{u}', t_{u'})|0\rangle\theta(t - t')$$

$$= \Sigma_{s,s'} \int d^{r1-1}k \int d^{r2-1}q \int d^{r1-1}k' \int d^{r2-1}q'\, \mathfrak{N}(q, k)\, \mathfrak{N}(q', k')\delta_{\beta\beta'}\,\delta_{\alpha\alpha'}\,\delta^{r1-1}(k - k')\, \delta^{r2-1}(q - q')\cdot$$

$$\cdot[e^{i(k\cdot x+q\cdot u)}e^{-i(k'\cdot x' + q'\cdot u')}\, u_{r1}(k', s')\overline{u}_{r1}(k, s)\theta(t' - t) +$$

$$+ e^{-i(k\cdot x+q\cdot u)}e^{i(k'\cdot x' + q'\cdot u')}\overline{v}_{r1}(k, s)v_{r1}(k', s')\,\theta(t - t')]$$

$$= \Sigma_s \int d^{r1-1}k \int d^{r2-1}q\, \mathfrak{N}(q, k)^2\delta_{\alpha\alpha'}\,\delta_{\beta\beta'}[e^{i(k\cdot(x - x') + q\cdot(u - u'))}\overline{u}_{r1}(k, s')u_{r1}(k, s)\theta(t' - t) +$$

$$+ e^{i(k\cdot(x' - x) + q\cdot(u' - u))}\, v_{r1}(k, s)\overline{v}_{r1}(k, s)\,\theta(t - t')]$$

$$= \int d^{r1-1}k \int d^{r2-1}q\ \mathfrak{N}(q, k)^2 \delta_{\alpha\alpha'} \delta_{\beta\beta'}[e^{i(k\cdot(x-x')+q\cdot(u-u'))}\ u_{r1}(k, s')\bar{u}_{r1}(k, s)\theta(t' - t) +$$

$$+ e^{i(k\cdot(x'-x)+q\cdot(u'-u))}\ v_{r1}(k, s)\bar{v}_{r1}(k, s')\ \theta(t - t')]$$

$$= \delta_{\alpha\alpha'} \delta_{\beta\beta'} \int d^{r1-1}k \int d^{r2-1}q\ (4k^0)^{-1}(2\pi)^{-(r1-1)-(r2-1)}\ [e^{i(k\cdot(x-x')+q\cdot(u-u'))}\ (\not{k} + m)\theta(t' - t) +$$

$$+ e^{-i(k\cdot(x'-x)+q\cdot(u'-u))}\ (m - \not{k})\theta(t - t')] \qquad (6.36)$$

If $u^0 = u^{0\prime}$ the universe part factors out, the inner universe is unchanged, and the host part is Feynman-like:

$$S_F(x' - x)\ \gamma^0 = \delta^{r2-1}(\mathbf{u} - \mathbf{u}')\ \delta_{\alpha\alpha'} \delta_{\beta\beta'} \int d^{r1-1}k\ (4k^0)^{-1}(2\pi)^{-(r1-1)}\ [e^{ik\cdot(x-x')}(\not{k} + m)\theta(t' - t) +$$

$$+ e^{ik\cdot(x'-x)}(m - \not{k})\theta(t - t')]$$

$$= \delta^{r2-1}(\mathbf{u} - \mathbf{u}')\ \delta_{\alpha\alpha'} \delta_{\beta\beta'} \int d^{r1}k(2\pi)^{-r1}\ e^{-ik\cdot(x'-x)}/(\not{k} - m + i\varepsilon) \qquad (6.37)$$

If $u^0 \neq u^{0\prime}$ then from eq. 6.13 we have

$$S_F(x' - x, u' - u)\gamma^0 = \delta_{\alpha\alpha'} \delta_{\beta\beta'} \int d^{r1-1}k \int d^{r2-1}q\ (4k^0)^{-1}(2\pi)^{-(r1-1)-(r2-1)}[e^{i(k\cdot(x-x')+q\cdot(u-u'))}(\not{k} + m)\theta(t' - t) +$$

$$+ e^{-i(k\cdot(x'-x)+q\cdot(u'-u))}\ (m - \not{k})\theta(t - t')] \qquad (6.38)$$

For $u^0 \neq u^{0\prime}$ we can express S_F in terms of the homogeneous wave equation solutions in r space-time dimensions:

$$\Delta_{rm}(y' - y) = -i \int d^{r-1}k\ (e^{-ik\cdot(y'-y)} - e^{ik\cdot(y'-y)})/[(2\pi)^{r-1}2\omega_k] \qquad (6.39)$$

$$\Delta_{1rm}(y' - y) = \int d^{r-1}k\ (e^{-ik\cdot(y'-y)} + e^{ik\cdot(y'-y)})/[(2\pi)^{r-1}2\omega_k] \qquad (6.40)$$

$$\Delta_{2rm}(y' - y) = \int d^{r-1}k\ e^{-ik\cdot(y'-y)}/[(2\pi)^{r-1}2\omega_k] = ½(\Delta_1(y' - y) + i\ \Delta(y' - y)) \qquad (6.41)$$

$$\Delta_{3rm}(y' - y) = \int d^{r-1}k\ e^{+ik\cdot(y'-y)}/[(2\pi)^{r-1}2\omega_k] = ½(\Delta_1(y' - y) - i\ \Delta(y' - y)) \qquad (6.42)$$

where $\omega_k = (\mathbf{k}^2 + m^2)^{½}$. Then

$$S_F(x' - x, u' - u)\gamma^0 = \delta_{\alpha\alpha'}\delta_{\beta\beta'}\{(-i\gamma\cdot\partial + m)\Delta_{3r1m}(x' - x)\partial/\partial u^0\Delta_{3r2M}(u' - u)\theta(t' - t) +$$

$$+ (-i\gamma\cdot\partial - m)\Delta_{2r1m}(x' - x)\partial/\partial u^0\Delta_{2r2M}(u' - u)\theta(t - t')\} \qquad (6.43)$$

The host fermion and the inner universe propagate "jointly" with an interaction between the Host propagation and the inner universe propagation.

A one host-universe state is created from the vacuum by

$$|N_1N_2\alpha\beta Mkqs> = b_{2N_1N_2\alpha\beta M}(k, q, s)^\dagger|0>_2 \quad (6.44)$$

using the vacuum $|0>_2$ of PseudoQuantum theory. The host-universe initial wave function is

$$\psi_{N_1N_2\alpha\beta M}(x, u, k, q, s) = <x, u \mid N_1N_2\alpha\beta Mkqs> \quad (6.45)$$

Note that a superposition of wave functions using a distribution of momenta, and mass M introduces a further quantum aspect in the initial universe particle state.

Interaction terms such as the electromagnetic-like interaction for atoms given earlier can be defined.

6.2 Free Host Space Boson with Host Internal Symmetries

This section is similar to section 6.1.2 but modified for bosons. It has an internal symmetry for a host boson in the host space-time. The symmetry will correspond to the host Blaha number N_1 space with $d_{dN_1} = 2^{22-2N_1}$ total dimensions by eq. 2.25. We introduce a corresponding host index β that ranges from 1 to $2^{22-2N_1} - r1$ where $r1 = 18 - 2N_1$ is the number of host space-time dimensions.

The Blaha number N_2 space (inner universe) has $d_{dN_2} = 2^{22-2N_2}$ total dimensions by eq. 2.25. We introduce a corresponding index α that ranges from 1 to $2^{22-2N_2} - r2$ where $r2 = 18 - 2N_2$ is the number of inner space-time dimensions.

The space and space-time dimensions of the host space and the inner universe are different in general. Either may have a greater space-time dimension.

We define the boson creation/annihilation operators with

$$a_{iN_1N_2\alpha\beta M'}(k, q)^\dagger \quad \text{and} \quad a_{i\,N_1N_2\alpha\beta M'}(k, q) \quad (6.46)$$

for PseudoQuantum i = 1, 2, M′ is the mass-energy of the created universe, N_1 is the host Blaha number, $r1 = 18 - 2N_1$ (by eq. 2.30)) is the number of space-time dimensions of the *host* universe with momentum $k = (k_1, k_2, \ldots, k_{r1})$. The mass of the host boson is m′.

Within the host boson we have an inner space (universe) of dimension $d_{dN_2} = 2^{22-2N_2}$, which contains an $r2 = 18 - 2N_2$ dimension real-valued space-time. The index α labels the dimensions of the inner N_2 space. We take $t_u = u_{r1}$ to be the time variable within the inner universe. The index α labels non-space-time components for the $d_{dN2} - r2$ real-valued dimension internal symmetry group representation for the inner universe.

The momentum k is an r1-vector for the host boson. The momentum q is an r2-vector for the embedded space (universe).

The non-zero anti-commutator relations are

$$[a_{iN_1N_2\alpha\beta M'}(k, q), a_{jN_1N_2\alpha'\beta'M'}(k', q')^\dagger] = \delta_{ss'}\delta_{\alpha\alpha'}\delta_{\beta'\beta}\delta(\mathbf{k} - \mathbf{k}')\,\delta^{r1-1}(\mathbf{q} - \mathbf{q}') \quad (6.47)$$

for i, j = 1, 2 with i ≠ j.

The free PseudoQuantum wave functions are:

$$\varphi_{iN_1N_2\alpha\beta M'}(\mathbf{x}, t, \mathbf{u}, t_u) = \int d^{r1-1}k \int d^{r2-1}q\ \mathfrak{N}(q, k)[a_{iN_1N_2\alpha\beta M'}(k, q)e^{-i(k\cdot x + q\cdot u)} + a_{iN_1N_2\alpha\beta M'}(k, q)^\dagger e^{i(k\cdot x + q\cdot u)}] \tag{6.48}$$

for i = 1, 2 where $\mathfrak{N}(q, k)$ is a normalization constant, where m′ is the host boson mass in the N_1 space's space-time and M′ is the mass-energy of the universe space of Blaha number N_2. We assume $k^2 = m'^2$ for the host boson and $q^2 = M'^2$ in the inner N_2 universe. Note $x = (x^0, \mathbf{x})$, $u = (u^0, \mathbf{u})$, $t = x^0$, and $t_u = u^0$. $\varphi_{iN_1N_2\alpha\beta M'}(x, u)$ forms a representation of $su(2^{9-N_2}, 2^{9-N_2})$ if one includes *both* internal symmetries and the inner universe's space-time. See Figs. 1.1 and 1.2.

We define the Lagrangian for a host fermion space particle interacting with a boson space particle containing a bosonic universe as:

$$\mathcal{L} = \overline{\psi}(x, u)[i\gamma^\mu\partial/\partial x^\mu - g'\varphi - m + (\partial/\partial u^\mu\ \partial/\partial u_\mu - M^2)/\lambda]\psi(x, u) +$$

$$+ ½(\partial/\partial x^\mu\varphi(x, u)\ \partial/\partial x_\mu\varphi(x, u)\ - m'^2\varphi^2(x, u) + (\partial/\partial u^\mu\ \varphi(x, u)\partial/\partial u_\mu\ \varphi(x, u) - M'^2\varphi^2(x, u)) \tag{6.49}$$

where m′ is the mass of the host φ space boson, and M′ is the mass of the inner φ space boson. The x host boson canonical momentum is

$$\pi_X(x, u) = \partial/\partial x^0\ \varphi(x, u) \tag{6.49a}$$

The separable free field boson dynamic equation is

$$[(\partial/\partial x^\mu\ \partial/\partial x_\mu - m'^2) + (\partial/\partial u^\mu\ \partial/\partial u_\mu - M^2)/\lambda]\ \varphi(x, u) = 0 \tag{6.50}$$

We separate this equation to obtain a host boson equation and an inner universe boson equation. The *host*, free field, equation is

$$(\partial/\partial x^\mu\ \partial/\partial x_\mu - m'^2)\varphi(x, u) = 0 \tag{6.51}$$

The inner space/universe of the boson also has a Klein-Gordan equation under the assumption that it is a particle.[45] Its equation is

$$(\partial/\partial u^\mu\ \partial/\partial u_\mu - M'^2)\varphi(x, u) = 0 \tag{6.52}$$

The two differential equations hold separately for the host and inner space coordinates.

The wave function equal time commutator is

$$[\pi_{XiN_1N_2\alpha\beta M'}(\mathbf{x}, t, \mathbf{u}, t_u), \varphi_{jN_1N_2\alpha'\beta'M'}(\mathbf{x}', t, \mathbf{u}', t_u)^\dagger] = \int d^{r1-1}k \int d^{r2-1}q \int d^{r1-1}k' \int d^{r2-1}q'\ (-ik^0)\mathfrak{N}(q, k)\ \mathfrak{N}(q', k')\cdot$$
$$\cdot[a_{iN_1N_2\alpha\beta M}(k, q)e^{-i(k\cdot x+q\cdot u)} - a_{iN_1N_2\alpha\beta M}{}^\dagger(k, q)e^{i(k\cdot x+q\cdot u)}], [a_{jN_1N_2\alpha'\beta'M}(k', q')^\dagger e^{i(k'\cdot x'+q'\cdot u')} + a_{jN_1N_2\alpha'\beta'M}(k', q')\cdot$$
$$\cdot e^{-i(k'\cdot x'+q'\cdot u')}]\}$$

---

[45] See Blaha (2021d) *Universes are Particles.*

$$= \int d^{r1-1}k \int d^{r2-1}q \int d^{r1-1}k' \int d^{r2-1}q' \, (-ik^0)\mathfrak{N}(q, k)\, \mathfrak{N}(q', k')\delta_{\alpha\alpha'}\delta_{\beta\beta'}\delta^{r1-1}(k - k')\, \delta^{r2-1}(q - q')\cdot$$

$$\cdot[e^{-i(k\cdot x+q\cdot u)}e^{i(k'\cdot x' + q'\cdot u')} - e^{i(k\cdot x+q\cdot u)}e^{-i(k'\cdot x' + q'\cdot u')}]$$

$$= \delta_{\alpha\alpha'}\delta_{\beta\beta'}\int d^{r1-1}k \int d^{r2-1}q\, (-ik^0)\, [e^{-i(k\cdot x+q\cdot u)}e^{i(k\cdot x' + q\cdot u')} - e^{i(k\cdot x+q\cdot u)}e^{-i(k\cdot x' + q\cdot u')}]\, \mathfrak{N}(q, k)^2$$

for $i \neq j$ where the space's time is t, the independent embedded universe's time is t_u, and γ^0 is a Dirac matrix for the host space-time. Note the spatial parts are in bold typeface. Note also $t \neq t_u$. They are independent variables. If

$$\mathfrak{N}(q, k) = (k^0)^{-\frac{1}{2}}(2\pi)^{-(r1-1)/2 - (r2-1)/2}$$

then

$$[\pi_{xiN_1N_2\alpha\beta M'}(\mathbf{x}, t, \mathbf{u}, t_u), \varphi_{jN_1N_2\alpha'\beta'M'}(\mathbf{x}', t, \mathbf{u}', t_u)^\dagger] = -i\delta_{\alpha\alpha'}\, \delta_{\beta\beta'}\delta^{r1-1}(\mathbf{x} - \mathbf{x}')\delta^{r2-1}(\mathbf{u} - \mathbf{u}') \qquad (6.53)$$

with other commutators zero.

One can show that the host boson and the inner universe (boson) propagate "jointly" with an interaction between the Host propagation and the inner universe propagation. There is some evidence that our universe does not have spin. Thus it is apparently bosonic.

6.3 Boson Host with a Three Fermion Inner Universes

We now consider the case of a universe with a boson host part, and having three fermion inner universes. In chapter 7 we suggest that it provides a model of a color confined hadron such as a proton.[46] The confinement mechanism is implemented at the level of the second quantization of the hadron with creation/annihilation operators—not with interaction terms in dynamical field equations.

Using the PseudoQuantum formulation we have the wave function:

$$\varphi_{iN_1N_2N_3N_4\alpha\beta\gamma\delta}(x, u_1, u_2, u_3) = \int d^{r-1}k \int d^{r1-1}q_1\, d^{r2-1}q_2 d^{r3-1}q_3 \mathfrak{N}(q_1, q_2, q_3, k)\cdot$$

$$\cdot\, [u_{r1}(q_1, s_1)u_{r2}(q_2, s_2)u_{r3}(q_3, s_3)b_{iN_1N_2N_3N_4\alpha\beta\gamma\delta}(k, q_1, q_2, q_3, s_1, s_2, s_3)e^{-i(k\cdot x + q_1\cdot u_1 + q_2\cdot u_2 + q_3\cdot u_3)} +$$

$$+ v_{r1}(q_1, s_1)v_{r2}(q_2, s_2)v_{r3}(q_3, s_3)d_{iN_1N_2N_3N_4\alpha\beta\gamma\delta}(k, q_1, q_2, q_3, s_1, s_2, s_3)^\dagger e^{i(k\cdot x + q_1\cdot u_1 + q_2\cdot u_2 + q_3\cdot u_3)}] \qquad (6.54)$$

for i = 1, 2 where $\mathfrak{N}$ is a normalization constant, where m is the host fermion mass in the N_1 space's space-time, M_1, M_2, and M_3 are the mass-energies of the three fermion universe spaces of Blaha number N_2, N_3 , and N_4 respectively. We assume $k^2 = m^2$ of the host boson and $q_j^2 = M_j^2$ in the inner spaces for j = 1, 2, 3.

[46] Quarks are spin ½ fermions. The host part must be a spin 0 boson since the composite proton is a spin ½ particle.

The s_1, s_2, s_3 are the inner fermion spins. When states are defined the spins may be combined to form a combined spin state for the host universe and the inner universes.

Note: a host space boson may have inner universes such as a universe – anti-universe pair of inner universes. Such a space boson could be used to model hadronic bosons made of quark-antiquark universes. See chapter 7.

6.4 Boson Host with n′ Fermion Inner Universes

The formalism may be expanded to create a host particle with any number of inner spaces. Thus one can reduce a scalar particle to a composite of "Gold Dust" and explore the consequence of a dust formulated particle. This approach may be used for fermions as well. The result is a host space fermion with up to an infinite number of inner (fermion or boson) universes.

A Lagrangian for a scalar space particle with n inner boson universes is:

$$\mathcal{L} = \bar{\psi}(x, u)[i\gamma^\mu\partial/\partial x^\mu - g'\varphi - m + (\partial/\partial u_j^{\,\mu}\,\partial/\partial u_{j\mu} - M_j^2)/\lambda]\psi(x, u) +$$

$$+ \tfrac{1}{2}(\partial/\partial x^\mu\varphi(x, u)\,\partial/\partial x_\mu\varphi(x, u) \;- m'^2\varphi^2(x, u) + \sum_{j=1}^{n} (\partial/\partial u_j^{\,\mu}\,\varphi(x, u)\partial/\partial u_{j\mu}\,\varphi(x, u) - M'^2_j\varphi^2(x, u))) \qquad (6.55)$$

In the limit $n \to \infty$ the boson space particle becomes a composite of "Gold Dust."

A Lagrangian for a fermion space particle with n inner boson universes is:

$$\mathcal{L} = \bar{\psi}(x, u)[i\gamma^\mu\partial/\partial x^\mu - g'\varphi - m + \sum_{j=1}^{n} (\partial/\partial u_j^{\,\mu}\,\partial/\partial u_{j\mu} - M_j^2)/\lambda]\psi(x, u) +$$

$$+ \tfrac{1}{2}(\partial/\partial x^\mu\varphi(x, u)\,\partial/\partial x_\mu\varphi(x, u) \;- m'^2\varphi^2(x, u) + (\partial/\partial u^\mu\,\varphi(x, u)\partial/\partial u_\mu\,\varphi(x, u) - M'^2\varphi^2(x, u))) \qquad (6.56)$$

A model Lagrangian for a scalar space particle with n inner fermion universes is:

$$\mathcal{L} = \bar{\psi}(x, u)[i\gamma^\mu\partial/\partial x^\mu - g'\varphi - m + \sum_{j=1}^{n'} (i\gamma^\mu\partial/\partial x_j^{\,\mu} - \; M_j^2)/\lambda]\psi(x, u) +$$

$$+ \tfrac{1}{2}(\partial/\partial x^\mu\varphi(x, u)\,\partial/\partial x_\mu\varphi(x, u) \;- m'^2\varphi^2(x, u) + \sum_{j=1}^{n} (\partial/\partial u_j^{\,\mu}\,\varphi(x, u)\partial/\partial u_{j\mu}\,\varphi(x, u) - M'^2_j\varphi^2(x, u))/\lambda) \qquad (6.55)$$

In the limit $n' \to \infty$ the boson space particle becomes a composite of fermionic "Gold Dust."[47] Eq. 6.54 illustrates the formalism for a space boson containing three inner fermion universes. For n′ inner fermions the wave function is:

$$\varphi_{iN_1N_2N_3N_4\alpha\beta\gamma\delta}(x, u_1, u_2, u_3) = \int d^{\,r-1}k \int\prod_{j=1}^{n'} d^{\,rj-1}q_j\; \mathfrak{N}(q_1, q_2, \ldots, q_{n'}, k)\cdot \qquad (6.57)$$

$$\cdot\,[\prod_{j=1}^{n'} u_{rj}(q_j, s_j)\; b_{iN_1N_2N_3\ldots N_{n'}\alpha_1\alpha_2\ldots\alpha_{n'}}(k, q_1, q_2, \ldots, q_{n'}, s_1, s_2, \ldots, s_{n'})e^{-i(k\cdot x + \Sigma\, q_j\cdot u_j)} +$$

[47] See Blaha (2022b).

$$+ \prod_{j=1}^{n'} v_{rj}(q_j, s_j) d_{iN_1N_2N_3 \ldots N_{n'}N_4\alpha_1\alpha_2\ldots\alpha_{n'}}(k, q_1, q_2, \ldots, q_{n'}, s_1, s_2, \ldots, s_{n'})^\dagger e^{i(k\cdot x + \Sigma q_j \cdot u_j)}]$$

where the subscript rj represents the jth space-time (r^{th}) dimension. Thus $u_{rj}(q_j, s_j)$ is the spinor for the jth fermion for the q_j^{th} space-time coordinates. As n′ grows the fineness of the "dust" becomes larger. The $n' = \infty$ limit is of interest.

7. Implementation of Particle Confinement with Space Wave Functions

The mechanism presented in chapter 6 enables the creation of a new type of particle "binding." We normally consider the binding of composite particles to take place through an interaction appearing in a Lagrangian. For example, this approach leads to the binding through interaction terms of a set of three quarks to form a proton. It could also be the basis of a new second quantized Bag Model.

The strong binding mechanism for quarks to form hadrons has the important issue of quark confinement. However the origin of the quark confinement is not decisively known. A linear potential or lattice-based mechanism is often viewed as its source.

A different binding mechanism, *space binding*, may be the result of the space particle formulation. Consider the case of a proton. If we assume a proton wave function embodies a boson host part containing three quark "universes", then we can treat the combined system as a form of bound state.[48] We presented a candidate space boson wave function in section 6.3.

Space binding directly implements quark confinement as we show later. For example a proton may transform to a different three quark state through interactions—but its nature as a three quark hadron is preserved after interactions. A similar comment applies to bosons consisting of two quarks.

Thus we can treat all hadrons as embodiments of space binding. The quarks (and anti-quarks) in the hadrons have a distinct identity. They are treated as independent particles although they can only reside within space particles.

Leptons do not enjoy confinement. They are free particles outside of interaction regions. We treat them as having "normal" wave functions.

We treat all vector bosons as "normal" vector bosons.

Thus the theory with quark confinement follows from the method by which wave functions are second quantized with creation and annihilation operators. The dynamical field equations are not the source of confinement. Note that space binding is consistent with baryon number conservation.

7.1 Deep Inelastic Lepton-Nucleon Scattering Case

The space binding formulation of deep inelastic lepton-nucleon scattering is illustrated by the Feynman-like diagram in Fig. 7.1. We find the interaction of the quark and incoming photon may result in a change in the quark identity. Yet the presence of the quark, or its successor, in the nucleon is guaranteed maintaining a three quark fermion and consequently color confinement.

---

[48] Quarks are spin ½ fermions. The host part must be a spin 0 boson since the composite proton is a spin ½ particle.

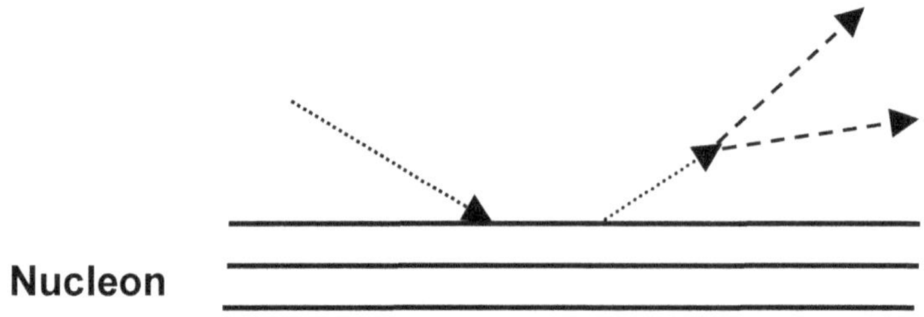

Figure 7.1. Feynman-like diagram for nucleon of three quark "universes" with one quark interacting with a virtual photon boson and emitting a vector boson that transforms into a pair of particles. The interacting quark may change its internal quantum numbers. Yet the nucleon continues to be composed of three quarks. A new form of Confinement!

7.2 Two Dimension Inner Fermions

The fermion particle described above may be considered to consist of a host boson in our four dimension space-time universe containing three inner quark universes. Each inner universe has a space-time dimension, which we could also choose to be four dimensions or otherwise.

However we might also consider the possibility that the inner quark universes are of Blaha number N = 8. The N = 8 spaces (universes) have two dimensions: one time dimension and one spatial dimension. They would correspond to a form of *string*. One could then develop a hybrid theory of hadrons composed of quark strings and leptons (normal quantum fields) and (normal) vector bosons. A new form of String theory!

7.3 Space Bag Theory

In the 1980's bag models were considered for hadronic fermions. A noteworthy bag model was created by an MIT group: The MIT Bag Model. The type of space particles described here lends itself as a new relativistic, second quantized formulation of Bag Theory. In this formulation a hadron is a *Space Bag* containing quarks and antiquarks. One defines a bag with a definite size as a superposition of mass-energy states. The inner fermions within the bag are also superposed. The resulting structure has a particle-like appearance. Inner quarks within a space bag interact with vector bosons in the usual way.

7.4 Multilevel Confinement

The binding mechanism described here may bind quarks within hadrons. Later we will consider the fractionization of particles and dimensions into "dust." After disassembling a particle in this manner we can view particles as bounded by a new form of confinement—as dust balls. The mechanism for this binding may be similar to the proposed quark binding mechanism—perhaps stronger—perhaps requiring energies far beyond those required to dissociate quarks (if that is possible)—perhaps only at energies seen at the creation of universes.

8. Generation of Universes by Fermion-Antifermion Annihilation

We now consider the generation of a sequence of spaces of the HyperCosmos from an ultimate space of Blaha space number 0 with 18 space-time dimensions. The generation proceeds through a fermion-antifermion annihilation in a "parent" space of level N with fermion wave function $\psi_{2N\alpha M}(x, u)$ that creates a pair of scalar space particles with each containing a "child" space of level N + 1 of mass-energy M_N. See Fig. 8.1. We assume the creation process is inelastic. It conserves spatial momentum in the parent space. Energy conservation includes a transfer of mass-energy to child spaces.

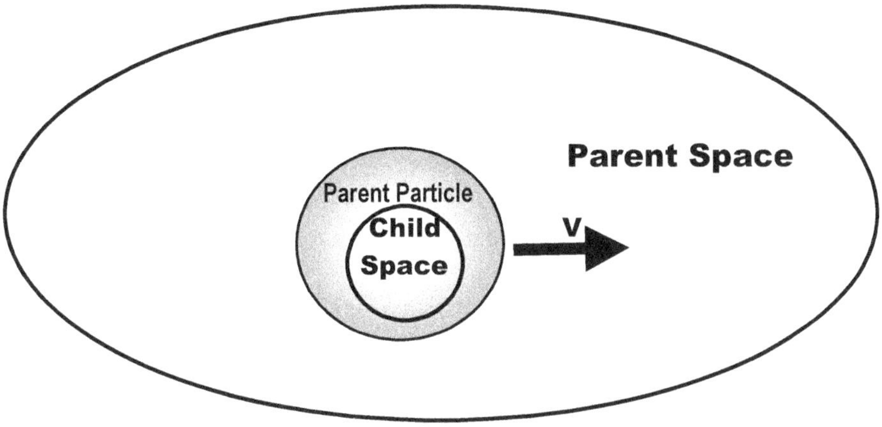

Figure 8.1. The general pattern of a (parent) space containing a parent particle of velocity **v** containing an internal space instance (child) of lower space-time dimension and higher Blaha number. The inner space is within an instance of a higher space. The spectrum's spaces form a nested sequence. Each space contains the next lower space within it. .

Following the notation of chapter 6 we define a space wave function in Blaha number N space (universe) with an inner space of Blaha number N: $\psi_{2,N,\alpha,M}(x, u)$. We also define a space boson in the Blaha space N universe with an inner space of Blaha number N + 1: $\varphi_{2,N+1,\gamma,M'}(x,u')$. We define an interaction term[49] using type 2 PseudoQuantum fields:

[49] In Two-Tier quantum field theory, which we implicitly assume, quartic interactions do not create divergences in perturbation theory.

$$g\overline{\psi}_{2,N,\alpha,M}(x,u)\psi_{2,N,\alpha,M}(x,u)\varphi_{2,N+1,\gamma,M'}(x,u')\varphi_{2,N+1,\gamma,M'}(x,u') \qquad (8.1)$$

with summations over α, and γ, where the x coordinates are space-time coordinates in the space N universe, the u coordinates are in the space-time of the fermion inner space with Blaha number N, and the u′ coordinates for the child's space-time of inner space N + 1.

By eq. 2.31 the α index of the child d_{dN+1} array ranges from 1 through 2^{9-N} while the index of the parent d_{dN} array ranges from 2^{11-N}, a factor of 4 different.

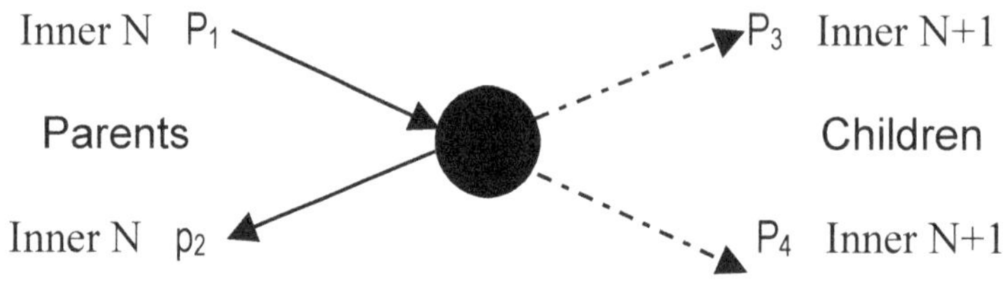

Figure 8.2. A Feynman-like diagram corresponding to the interaction in eq. 8.1.

It appears that the eq. 8.1 interaction is necessarily quadratic in both the fermion and boson fields to achieve the transition from space N to space N + 1. Thus fermion parent annihilation in the parent universe of Number N, and two child universes, within the boson fields, of number N + 1 are created in the parent universe of Number N

Thus the creation process creates a nested sequence of parents and children. See Fig. 8.3.

Note: the parent and child space-time dimensions in the eq. 8.1 interaction are related by

$$r_{parent} = r_{child} + 2 \qquad (8.2)$$

consistent with the creation of a space in the N + 1 level of the HyperCosmos spectrum of spaces.

We may take the parent inner universe masses $M_1 = M_2 = 0$ initially. The interaction may be inelastic with a transfer of mass-energy to the scalar particle inner universes (spaces) with masses M_3 and M_4 non-zero in general. We view the interaction as analogous to colliding "mud balls" where energy of motion is transferred to the resultant mud balls.

The spatial momentum is conserved:

$$\mathbf{P}_1 + \mathbf{P}_2 = \mathbf{P}_3 + \mathbf{P}_4 \qquad (8.3)$$

Energy is conserved if account is taken of the transfer of energy to the inner universe (space) within each mud ball. With the energy transfer the generated spaces can then evolve by expansion as we see in our own universe.[50]

$$E_1 + M_1 + E_2 + M_2 = E_3 + M_3 + E_4 + M_4 \qquad (8.4)$$

8.1 HyperCosmos Spectrum

Starting from the Blaha number N = 0 space with 18 space-time dimensions we can generate the remainder of the 10 spaces through fermion-antifermion annihilation processes as depicted in Fig. 8.3. We assume the N = 0 space is a universe containing mass-energy that supports the creation process. (The N = 0 universe (space) is generated in the ProtoCosmos.)

The sequence of spaces is nested. There may be several chains of spaces due to the stage by stage generation of sibling spaces. Fig. 5.2 shows a hierarchy of spaces generated through fermion-antifermion annihilation.

8.2 Sibling Universes

The creation of sibling universes may account for some unusual asymmetries in our universe such as the observed excess of Dark matter over Normal matter.

---

[50] See Blaha (2021d).

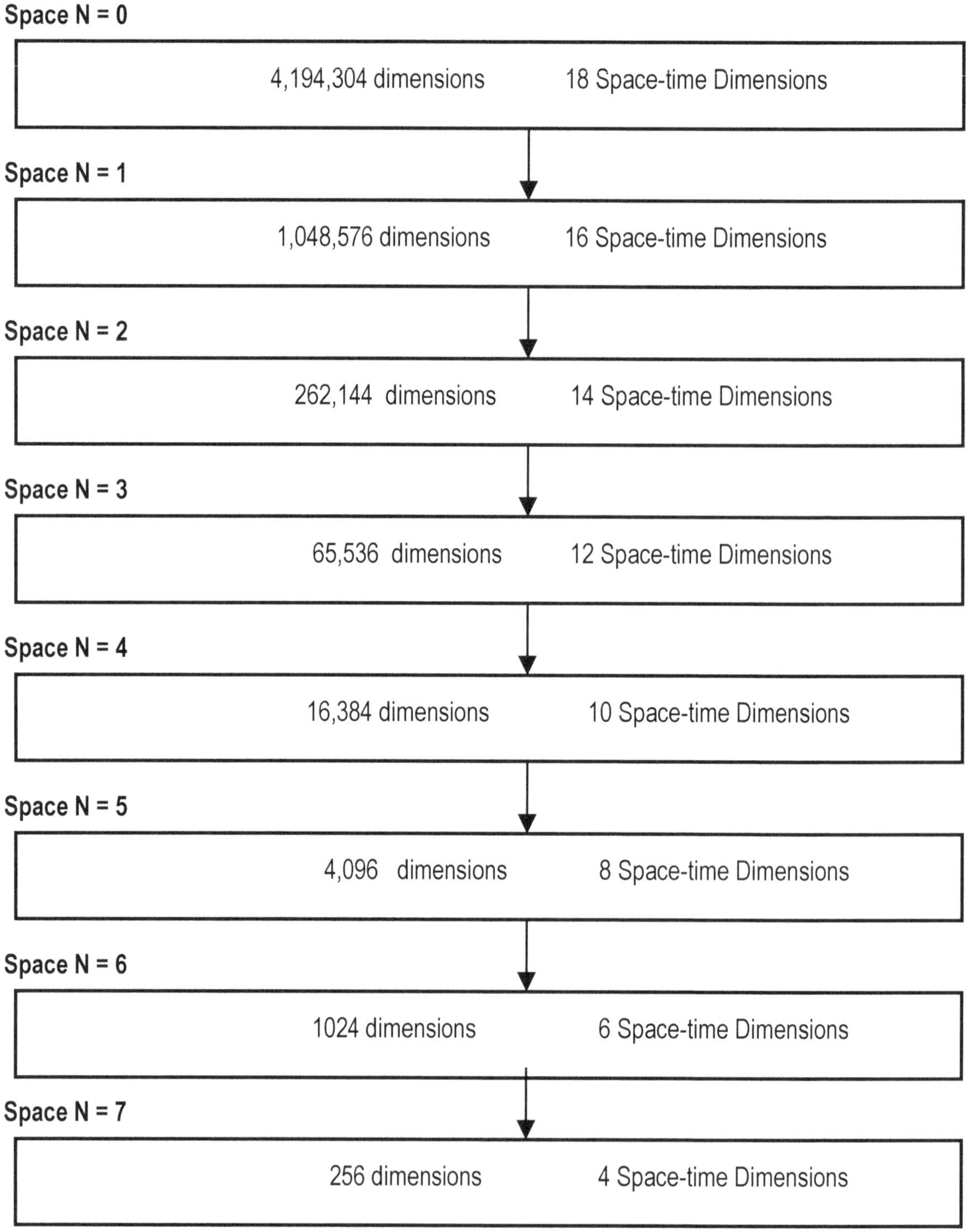

Figure 8.3. A sequence of fermion-antifermion annihilations generating instances in the set of HyperCosmos spaces leading to our universe of N = 7.

9. Fundamental Reference Frames

A common feature of calculations in Physics is to choose a reference frame which facilitates computation. Common choices are the rest frame and the center of mass frame. In this chapter we find a Fundamental Reference Frame (actually an infinite set of such frames) that plays a similar role for coordinate reference frames. We describe how to determine a Fundamental Reference Frame for each of the ten spaces (universes) of the HyperCosmos. We are motivated in this endeavor by the observation made in previous books (and later) that the fermion particles of a space (universe) appear to be expressible as sets of repeated fermions. We also see a similar duplication in the set of symmetries of a space. They are also expressible as sets of repeated products of symmetries. Sections 9.1 and 9.2 show this for an N = 7 space, which we have considered in our discussions of QUeST and the Unified SuperStandard Theory (UST). Remarkably, the set of duplicated fermions is the Standard Model fermions with an additional fourth generation. And the set of duplicated symmetries is $SU(3) \otimes U(1) \otimes SU(2) \otimes U(1) \otimes SO^{+}(1, 3)$—those of the Standard Model. *Thus we show that the Fundamental Reference Frame is based on a Standard Model set of fermions and symmetries.*

Based on these observations we generalize the sets of creation/annihilation operators seen in chapters 4 – 6 to support General Relativistic transformations that map a core set of fermions and of internal symmetries in a Fundamental Reference Frame to sets of repeated multiples of fermions, and of repeated sets of internal symmetries in a conventional reference frame. The key to this generalization is the use of hypercomplex formulations of creation/annihilation operators. They each become a hypercomplex-valued operator.

9.1 Set of Fermions Expressed as Repeated Sets of Fermions in QUeST and UST

The fundamental fermion spectrum was derived in Blaha (2018e) (and earlier books) from symmetry groups that followed from (broken) fermion conservation laws. It consisted of 256 fermions arranged in four generations in four layers. of "Normal" fermions and an identical spectrum of Dark fermions. Remarkably this spectrum was surprisingly, and encouragingly, found in hypercomplex spaces studies by this author in 2019-2022. See the resulting book: Blaha (2020a) Since then the author has developed hypercomplex spaces theories eventually resulting in the HyperCosmos (HyperComplex Cosmology) found recently.

The form of the 256 fermion spectrum for our universe (N = 7) continues to hold in the HyperCosmos. It may be viewed as 16 duplicates of a set of 16 Normal and Dark fermions of the species: electron, neutrino, three up-type quarks, and three down-type quarks. Each of 16 fermion sets has its own set of quantum numbers and its own set of interactions. We are familiar with the level 1, three generations of The Standard Model, which is incorporated within the UST. The UST and HyperCosmos suggest *four* generations.

The 16-fold repetition of the 16 fermion sets leads us to consider the possibility that they are the result of a duplication process. We found a formulation of the duplication process based on General Relativity applied to creation/annihilation operators in a Fundamental Reference Frame discussed later. The formulation applies to all spaces of the HyperCosmos. All spaces of the HyperCosmos exhibit patterns of repetition.. Thus the duplication process necessarily applies to them as well.

	NORMAL		DARK	
	e q-up	**v q-down**	**e q-up**	**v q-down**
	e q-up	**v q-down**	**e q-up**	**v q-down**
Layer 1	**e q-up**	**v q-down**	**e q-up**	**v q-down**
	e q-up	**v q-down**	**e q-up**	**v q-down**
	e q-up	**v q-down**	**e q-up**	**v q-down**
	e q-up	**v q-down**	**e q-up**	**v q-down**
Layer 2	**e q-up**	**v q-down**	**e q-up**	**v q-down**
	e q-up	**v q-down**	**e q-up**	**v q-down**
	e q-up	**v q-down**	**e q-up**	**v q-down**
	e q-up	**v q-down**	**e q-up**	**v q-down**
Layer 3	**e q-up**	**v q-down**	**e q-up**	**v q-down**
	e q-up	**v q-down**	**e q-up**	**v q-down**
	e q-up	**v q-down**	**e q-up**	**v q-down**
	e q-up	**v q-down**	**e q-up**	**v q-down**
Layer 4	**e q-up**	**v q-down**	**e q-up**	**v q-down**
	e q-up	**v q-down**	**e q-up**	**v q-down**

Figure 9.1. Four layers of fermions with each layer containing four generations. The fermions are arranged in SU(4)-like sets of four fermions based on the structures specified in chapter 10. The sets could be arranged to a different grouping without changing the discussion in this chapter.

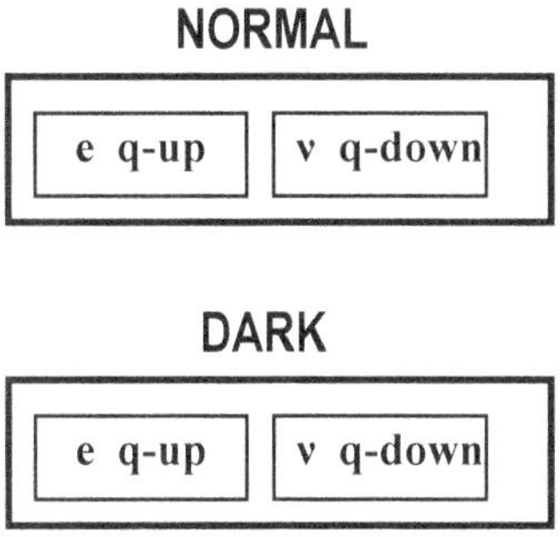

Figure 9.2. The 16 fermions that constitute the set that is duplicated 16 times in the 256 fermion spectrum for N = 7 space (universe).

The separation of the 16 fermions into pairs of four fermions in Fig. 9.2 and the form of Fig. 9.1 reflects the separation of fermions and internal symmetries into 4, 8, 16 32, and 64 parts due to symmetry splitting as we discussed in Blaha (2021b). We return to the discussion of symmetry splitting in chapter 15.

9.2 Set of Symmetry Groups Expressed as Repeated Sets of Groups in QUeST and UST

The set of symmetry groups in QUeST for our universe also exhibits the same sort of duplication seen for fundamental fermions. The symmetry groups have a 256 real-valued dimensions representation as the product of the fundamental representations of their component groups. The basic symmetry group set for an N = 7 space is a U(4)⊗U(4) group. After a 16-fold duplication the U(4) factors undergo transformations/breakdowns to

U(4) or SU(3)⊗U(1) or SU(2)⊗U(1)⊗SL(2, **C**)[51]

Initially the N = 7 full set of groups has the form of 16 U(4)⊗U(4) symmetry groups (Fig. 9.3.) Then they are transformed by symmetry breaking to the groups in Fig. 9.4.

[51] We use SL(2, C) to represent $SO^+(1, 3)$.

	NORMAL	DARK
Layer 1	U(4)⊗U(4) U(4)⊗U(4)	U(4)⊗U(4) U(4)⊗U(4)
Layer 2	U(4)⊗U(4) U(4)⊗U(4)	U(4)⊗U(4) U(4)⊗U(4)
Layer 3	U(4)⊗U(4) U(4)⊗U(4)	U(4)⊗U(4) U(4)⊗U(4)
Layer 4	U(4)⊗U(4) U(4)⊗U(4)	U(4)⊗U(4) U(4)⊗U(4)

Figure 9.3. The "initial" distribution of sets of N = 7 symmetry groups. Each set is distinct and supports interactions only for the corresponding set of fermions (separately for Normal and Dark fermions) in Fig. 9.1. *Thus each set of 16 fermion generations has its own quantum numbers and interactions.* Each U(4)⊗U(4) set has a 16 real-valued dimension representation, which will be of importance when we consider Fundamental Reference Frames in section 9.3.

NORMAL		DARK	
SU(3)⊗U(1) Generation U(4)	SU(2)⊗U(1)⊗SL(2, C) Layer U(4)	SU(3)⊗U(1) Generation U(4)	SU(2)⊗U(1)⊗SL(2, C) Layer U(4)
SU(3)⊗U(1) Generation U(4)	SU(2)⊗U(1)⊗SL(2, C) Layer U(4)	SU(3)⊗U(1) Generation U(4)	SU(2)⊗U(1)⊗SL(2, C) Layer U(4)
SU(3)⊗U(1) Generation U(4)	SU(2)⊗U(1)⊗SL(2, C) Layer U(4)	SU(3)⊗U(1) Generation U(4)	SU(2)⊗U(1)⊗SL(2, C) Layer U(4)
SU(3)⊗U(1) Generation U(4)	SU(2)⊗U(1)⊗SL(2, C) Layer U(4)	SU(3)⊗U(1) Generation U(4)	SU(2)⊗U(1)⊗SL(2, C) Layer U(4)

Figure 9.4. The transformed/broken sets of symmetries in QUeST (UST) and in N = 7 HyperCosmos space.. Note each element has a 16 real dimension representation. This depiction is also evident in QUeST and the UST. The SL(2, **C**) representation has four coordinates.[52]

9.2.1 Are there Dynamics in the Fundamental Reference Frame?

Since the Fundamental Reference Frame has symmetry groups of the form U(4)⊗U(4) there is no time in it[53]—no dynamical evolution. After the transformation[54] to a static reference frame, SL(2, C) is generated with time evolution as a result.

---

[52] The Lorentz Group $SO^+(1, 3)$ is often specified with an SL(2, **C**) representation.

[53] It is a non-static frame with no time defined.

[54] The transformation from the Fundamental Reference Frame to a static frame might be designed to embody symmetry breaking with Fig. 9.4 the result.

9.3 General Relativistic Transformation from a Fundamental Reference Frame to a Static Coordinate System

We now develop a formulation for the transformation of a core set of fermions and symmetry groups to sets of duplicates such as those described in sections 9.1 and 9.2. The formulation will apply to all HyperCosmos spaces (universes). This formulation was first presented in Blaha (2021j).

We begin by generalizing the sets of creation/annihilation operators of chapters 4 – 8 to support General Relativistic transformations that map a core set of fermions and of internal symmetries in a Fundamental Reference Frame to sets of repeated multiples of fermions and of repeated sets of internal symmetries in a conventional static reference frame. The key to this generalization is the use of hypercomplex extensions of creation/annihilation operators. Each normal creation/annihilation operator becomes a hypercomplex operator.

We will consider the general case of the space and universes of Blaha number N. (The particular case of Blaha number 7 (our space and universe) was considered in some detail above and in Blaha (2021j).

The number of elements in the symmetry array d_{dN} for Blaha number N is

$$d_{dN} = 2^{22-2N} = 2^{r+4} \tag{4.67}$$

where r is the space-time dimension. It satisfies

$$N = Os = ½(18 - r) \tag{2.30}$$

The symmetry array is a square array whose rows and columns have 2^{11-N} elements. The associated CASe group is su($2^{r/2}$, $2^{r/2}$). Its representation, which is composed of the direct product of fundamental representations, has a real-valued vector w_N with 2^{11-N} elements. We can also define vector v_N with the $d_{dN} = 2^{22-2N}$ elements of the dimension array. Note that the elements of v_N can be placed in a one-to-one correspondence with the b's and d's of the fermion with the least half-integer spin in the space of Blaha number N. The vector v_N can be placed in a one-to-one correspondence with the set of fundamental fermions of the space, and with the set of dimensions of the fundamental representations of the symmetries of the space (including space-time dimensions.)

Now we note that General Relativistic transformations may take an individual b or d creation/annihilation operator and transform it using a CASe group su($2^{r/2}$, $2^{r/2}$) transformation with a 2^{11-N} by 2^{11-N} rows/columns matrix representation to a sum of creation/annihilation operators for each momentum.

We take this concept and use it to relate the vectors w_N to the vectors w_N to v_N. To do so, we must *initially* use a reducible representation for su($2^{r/2}$, $2^{r/2}$) with a matrix representation with a $2^{22-2N} = 2^{r+4}$ by $2^{22-2N} = 2^{r+4}$ rows/columns. The vector v_N has 2^{22-2N} components. However w_N has only 2^{11-N} components.

We therefore define an enhanced vector W_N with 2^{22-2N} components by mapping each element of w_N from a real-valued number to a hypercomplex number with 2^{11-N} elements. The hypercomplex number may be viewed as composed of one non-zero element followed by $2^{11-N} - 1$ elements containing zeroes. The transformation of eq. 9.1 creates a new vector v_N with all components non-zero in general. This vector is a static space-time coordinate vector for fermions or symmetries or the dimensions of the dimension array d_{dN}.

We augment the General Relativistic transformation by generalizing su($2^{r/2}$, $2^{r/2}$) to su(2^r, 2^r). It has a matrix representation with $2^{22-2N} = 2^{r+4}$ by $2^{22-2N} = 2^{r+4}$ rows and columns. Thus we have the transformation

$$v_N = [T]\, W_N \qquad (9.1)$$

where T has a $2^{22-2N} = 2^{r+4}$ by $2^{22-2N} = 2^{r+4}$ matrix form.

We may take the vector space of the W_N to be a non-static Fundamental Reference Frame since almost all reference frames are non-static.

The group su(2^r, 2^r) is beyond the General Relativistic-based group su($2^{r/2}$, $2^{r/2}$). By augmenting the reducible su($2^{r/2}$, $2^{r/2}$) representation to make an irreducible su(2^r,2^r) representation, we require transformations from the Fundamental Reference Frame to a static frame to include General Relativity ***and*** *symmetry group aspects. The su(2^r,2^r) group unites General Relativistic transformations with internal symmetry transformations.* The study of its detailed form appears to be a worthwhile endeavor.

9.4 Contents of the Fundamental Reference Frame

The contents of a Fundamental Reference Frame number 2^{11-N} fermions or dimensions or symmetry dimensions (for each symmetry's fundamental representation.) In each case these items are distributed within the hypercomplex coordinates of the frame. For example, for N = 7 there are 16 fermions residing within 16 hypercomplex (sedenion) coordinates with the vector W_7 thus containing 256 real-valued coordinates.

The number of physical elements in the Fundamental Reference Frame of space N is 2^{11-N} and the total number of elements in the hypercomplex vector containing them is 2^{22-2N}. A transformation of the type of eq. 9.1 causes the generation of 2^{22-2N} elements representing duplication of fermions, internal symmetry dimensions or dimensions in general. Thus the transformation effectively "squares" the number of physically relevant quantities.

One can view a Fundamental Reference Frame as the "rest" frame for a HyperCosmos space. A generalized General Relativistic transformation maps the "rest" frame to a static coordinate system with duplicates of the contents of the rest frame.

Fig. 9.5 below tabulates the contents of the ten Fundamental Reference Frames of the ten HyperCosmos spaces (universes).

Contents of Fundamental Reference Frames

N	Number of Physical Elements	Total Number of Elements
0	2^{11}	2^{22}
1	2^{10}	2^{20}
2	2^{9}	2^{18}
3	2^{8}	2^{16}
4	2^{7}	2^{14}
5	2^{6}	2^{12}
6	2^{5}	2^{10}
7	2^{4}	2^{8}
8	2^{3}	2^{6}
9	2^{2}	2^{4}

Figure 9.5. The contents of the ten types of Fundamental Reference Frames for the ten spaces. The elements are fermions or dimensions or symmetry fundamental representation dimensions.

The contents of a Fundamental Reference Frame are fermions or symmetries or dimensions. We display the contents for the case of fermions for $N > 5$ in Fig. 9.6. Note the $N = 9$ space has units of fermions composed of four fermions. The number of units repeatedly doubles as we ascend up to $N = 0$.

9.5 Contents of the Static Frame

The static reference frame generated by the transformation in eq. 9.1 has 2^{11-N} copies of the 2^{11-N} items in the Fundamental Reference Frame. For example for $N = 7$ (our universe) 16 copies are made of the 16 items in its Fundamental Reference Frame as seen earlier. The copies are distributed in four layers of four generations of Normal and Dark fundamental fermions (or symmetry dimensions). Thus the 2^{22-2N} elements of the dimension array (or the fermion array obtained from it) consist of 2^{11-N} copies of the 2^{11-N} physical elements in the Fundamental Reference Frame.

Contents of a Fundamental Reference Frame for Fermions

N	Number of Physical Elements	Fermion Content	
0	2^{11}		
1	2^{10}		
2	2^{9}		
3	2^{8}	•	
4	2^{7}	•	
5	2^{6}	•	
6	2^{5}	NORMAL e q-up v q-down e′ q-up′ v′ q-down′ DARK e q-up v q-down e″ q-up″ v″ q-down″	Total: 32 fermions
7	2^{4}	Normal: e q-up v q-down Dark: e q-up v q-down	16 fermions
8	2^{3}	e q-up AND v q-down	8 fermions
9	2^{2}	e q-up OR v q-down	4 fermions

Figure 9.6. Possible contents of five of the ten types of Fundamental Reference Frames for fermions. The primes distinguish different fermions.

The symmetry group contents in the Fundamental Reference Frame are listed for N > 5 in Fig. 9.7. Note the units of symmetry irreducible representations dimensions are four real-valued dimensions. Irreducible symmetry group units also have four dimensions. The number of units repeatedly doubles as we ascend up to N = 0.

Contents of a Fundamental Reference Frame for Symmetry Group Dimensions

N	Number of Physical Elements	Fermion Content	
0	2^{11}		
1	2^{10}		
2	2^{9}		
3	2^{8}	•	
4	2^{7}	•	
5	2^{6}	•	
6	2^{5}	NORMAL SU(2)⊗U(1)⊗SU(4)⊗SL(2,C) DARK SU(2)⊗U(1)⊗SU(4)⊗SL(2,C)	Total: 32 dimensions
7	2^{4}	SU(2)⊗U(1)⊗SL(1,C)⊗ SU(4)	16 dimensions
8	2^{3}	SU(3)⊗U(1) or SU(4)	8 dimensions
9	2^{2}	SU(2)⊗U(1)	4 dimensions

Figure 9.7. Possible contents of five of the ten types of Fundamental Reference Frames for symmetry dimensions. The groups may undergo a further breakdown after being mapped to a static space-time to give terms like those in Fig. 9.4.

We describe the generation of the form of symmetries in chapters 10 and 15.

9.6 Meaning of Space-Time in the HyperCosmos

Each space-time within the HyperCosmos spectrum of spaces plays an important role. The question arises: What sets the number of dimensions in a space-time? In this section we answer that question and show a direct connection to the Fundamental Reference Frames. We begin by noting

$$r = 18 - 2N \qquad (9.2)$$
$$d_{dN} = 2^{22-2N} = 2^{r+4}$$

where r is the dimension of the space-time, and d_{dN} is the number of elements in the dimension array for Blaha space N. These equations imply

$$r = \log_2 (d_{dN}/16) \qquad (9.3)$$

Noting that d_{dN} is the square of the number of physical elements d_{FRF} in the Fundamental Reference Frame.:

$$d_{dN} = d_{FRF}^{\,2} \tag{9.4}$$

we see

$$r = 2 \log_2 (d_{FRF}/4) \tag{9.5}$$

or, in words, r is the logarithm (base 2) of the total number of copies of the set of d_{dN} fermions in a static frame divided by two, and twice the logarithm (base 2) of the total number of units of four elements of physical fermions or symmetries or dimensions in the non-static Fundamental Reference Frame. Note the dimensions in the dimension array d_{dN} include the dimensions of all the internal symmetry irreducible representations and the space-time dimensions.

Thus the space-time dimension of each space may be viewed as determined ultimately by its Fundamental Reference Frame physical elements. The number of space-time dimensions in a space, the total number of internal symmetry dimensions of a space, and the number of fundamental fermions of the space's Fundamental Frame are linked thus supporting the reality of the Fundamental Reference Frame.

This chapter has shown how the sets of fermions and dimensions are generated by a CASe transformation due to a General Relativistic transformation of the fermions and dimensions from the Fundamental Reference Frame to a static reference frame

9.7 A Physical Implementation of Platonic Realism: Plato's Theory of Abstract Forms

Platonic Realism views the universe as based on universals or abstract objects, which are called Ideal Forms. The world (universe) is described as imperfect dynamic embodiments of these unchanging abstract forms. The Fundamental Reference Frames that we have found may be viewed as implementing the ideal forms of Plato. A Fundamental Reference Frame has fermions, dimensions and symmetries defined as abstract entities. They are fixed, and unchanging, since time and interactions are not proceeding dynamically. Fundamental Reference Frames are non-static reference frames with no defined time evolution.

HyperCosmos spaces are defined for static reference frames. They are the dynamically changing, imperfect embodiments of the Fundamental Reference frames "Ideal Forms."

Each of the ten spaces of the Hyper Cosmos has a set of Fundamental Reference Frames. The contents of the frames show a progression in the sense that the contents of the reference frame of the space of Balham number N is four times the contents of the reference frame of space N + 1. The four-fold increase in contents for space N is a four-fold duplication of the contents for the space N + 1.

The requirements for a physical analogue of Platonic theory of Realism therefore appear to be met in a physical context by Hyper Cosmos spaces and their Fundamental Reference Frames.

10. Space-Time Determination of Fermion Particle Species

We now turn to consider the form of fermion species due to the appearance of two times in su(1, 1).[55] The su(1, 1) CASe group determines the form of the species of fermions in all HyperCosmos spaces since it is present in all spaces in the simplest boson wave functions.

The su(1, 1) group has a metric with two real time coordinates:

$$ds^2 = t_{01}^{\ 2} + t_{02}^{\ 2} - x_1^{\ 2} - x_2^{\ 2} \tag{10.1}$$

In Blaha (2007b) we showed that the four species of fermions: e-type, ν-type, q-up-type and q-down-type followed from the four types of boosts of the Lorentz Group $SO^+(1, 3)$, which is often expressed as SL(2, **C**)). Complex Lorentz group boosts separate into sublight, superluminal, complex sublight, and complex superluminal boosts yielding the four species respectively.

Now we have a more interesting situation with two time coordinates.[56] Each has its own light speed. We take them to be numerically equal for the purposes of our discussion. One light speed divides the set of possible fermions into two "superspecies", namely normal and Dark fermions. The second light speed further divides each superspecies into four parts. The result is the eight fermion species *in our universe, Fig. 10.1. It* includes both Normal and Dark sectors. It is also the form of fermion species in the other HyperCosmos spaces which also have su(1, 1) multiple time coordinates.

The 16 fermions of an N = 7 Fundamental Reference Frame can be separated into the eight subspecies if account is taken of the occurrence of each quark species as a triplet. This separation of fermion species leads to the structure of the Unified SuperStandard Theory (UST). It also leads to the form of Fundamental Reference Frames for all HyperCosmos spaces as described in chapter 9.

A complete, detailed spectrum of fundamental fermions in our universe, and other universes, thus emerges, part of which appears in the Standard Model for our universe.

10.1 The Four Fermion Species Becomes Eight Subspecies

In Blaha (2007b) we showed that four species of fermions: e-type, ν-type, q-up-type and q-down-type followed from the four types of boosts of the Lorentz Group $SO^+(1, 3)$, which is often expressed as SL(2, **C**)). Complex Lorentz group boosts

[55] See chapter 4.

[56] Chapter 9 describes the transformation between a Fundamental Reference Frame where superluminal quantities are present and a static frame where they are transformed to sublight quantities.

separate into sublight, superluminal, complex sublight, and complex superluminal boosts yielding the four species respectively.

t_{01}:	sublight Normal				superluminal Dark			
t_{02}:	sublight		superluminal		sublight		superluminal	
	e-type	q-up-type	ν-type	q-down-type	e-type	q-up-type	ν-type	q-down-type

Figure 10.1. Separation of fermion species due to su(1, 1) and its two times. The total number of corresponding fermions is 16 if account is taken of quarks appearing as triplets. The occurrence of quark in triplets in the HyperCosmos is motivated by the need for fermions appearing in multiples of four units. Thus the separation by two light speeds may account for the Standard Model form of the fermion species and also for triplets of quarks. Experimentally, superluminal neutrinos appear to move at the speed of light due to their negligible masses. Down-type quarks, being confined within particles, do not directly display their superluminal nature. Dark matter, being unobserved may well be superluminal. Their superluminal nature may be part of the cause for their current unobservability. It is also possible that the superluminal nature of these quantities may "wash out" when a transformation is made from non-static to static reference frames.

10.2 Symmetries Separation Due to Space-Time

The symmetries in the Fundamental Reference Frame also have representations comprised of irreducible group representations with totals equal to multiple units of 4 real dimensions. Again the su(1, 1) CASe group with its two time dimensions structures the set of groups. Fig. 10.2 displays the separation of the representations due to sublight vs. superluminal parts.

10.3 Fundamental Reference Frames Contents

In chapter 9 we constructed Fundamental Reference Frames. Their units' contents for fermions and symmetries are structured according to the formatting presented in this chapter. Thus we have a formulation for the structure of the Fundamental Reference Frame and static space-time.

10.4 Total View of HyperCosmos Symmetries

The reduction of *each* HyperCosmos space to a Fundamental Reference Frame set of symmetries and fermions together with the partitioning of fermions and symmetries seen above in this chapter suggests a ***complete HyperCosmos structuring of the fermion spectrum and of the set of symmetries irrespective of the details of symmetry breaking and irrespective of complex proposals for internal symmetries such as SU(3)⊗SU(3).***

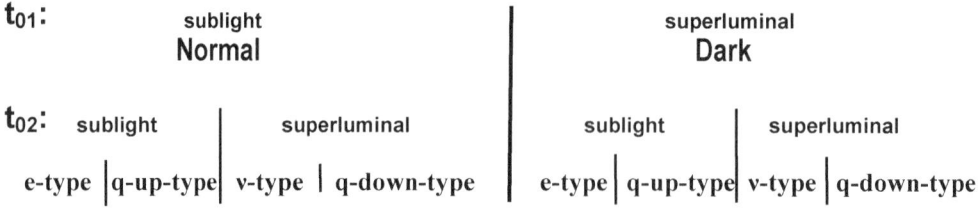

t_{01}:	sublight		superluminal	
	U(4)		U(4)	
t_{02}:	sublight	superluminal	sublight	superluminal
	$SO^+(1, 3)$[57]	$SU(2)\otimes U(1)$	U(1)	SU(3)
Representation Real Dimensions	4	4	2	6

Figure 10.2. Separation of symmetries due to su(1, 1) and its two times. The units' number of dimensions appears as muultiples of 4. The separation by two light speeds may account for the Standard Model form of the interaction groups. Superluminal SU(3) may account, in part, for color confinement, which remains somewhat of a mystery. The superluminal symmetries may appear las sublight symmetries in experiments. The $SU(2)\otimes U(1)$ symmetry superluminality is submerged by the extremely small neutrino masses. It is also possible that the superluminal nature of these quantities may "wash out" when a transformation is made from non-static to static reference frames. In Blaha (2018e), and earlier books, we constructed $SU(2)\otimes U(1)$ models based on complex space-time.

[57] For higher space-time dimension spaces the $SO^+(1, 3)$ copies transform to Connection Groups and the space's space-time dimension. See chapter 11.

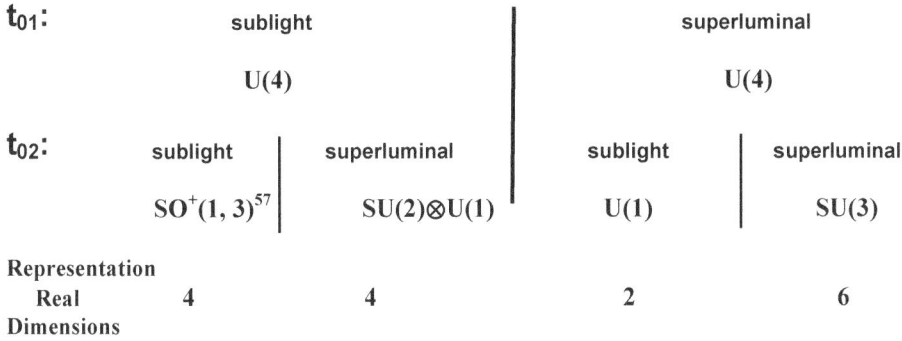

11. Hypercomplex Space-Time Coordinates and Connection Group Symmetries

Hypercomplex numbers are known to be related to symmetry groups. In this section we consider the multiple space-time symmetries that appear in the separation of dimension arrays into representations of groups. We have suggested in Blaha (2021b), (2021e) and (2021g) that the set of hypercomplex space-time dimensions be transformed to a set symmetry groups in each HyperCosmos space.

There is a two-fold justification for this procedure: 1). There is no evidence for the existence of hypercomplex numbers in our universe. 2) The generated set of symmetry groups has the important purpose of providing ultra-weak interactions uniting Normal and Dark matter. Without unification, Dark matter becomes physically irrelevant fot elementary particle physics.

We begin by noting the space-time dimension is set by the dimension array:

$$r = \log_2 (d_{dN}/16) \tag{9.3}$$

For N = 9 the space-time dimension is 0 since $d_{dN} = 16$. For N = 7 (our universe's space) the space-time dimension is 4 since $d_{dN} = 256$. We will consider the cases of NEWQUeST, NEWUTMOST, and NEWMaxiverse to illustrate the method.

11.1 Hypercomplex Coordinates Transformed to Symmetry Groups in Our Universe

Our QUeST formulation for our N = 7 universe has a $d_{d7} = 256$ component dimension array. A preliminary view of the symmetry groups that it implies appears in Fig. 11.1. Note that there are 32 dimensions that are initially allocated to space-time suggesting a hypercomplex space-time. We chose to allocate 4 real dimensions to obtain the 4-dimension space-time implied by eq. 9.3. The remaining 28 real dimensions we allocated to additional symmetry groups—namely seven U(2) groups. We call these groups Connection groups since their role is to "connect" fermions residing in different fermion spectrum blocks.

The structure of the seven additional U(2) groups is not specified. We chose to use a reasonable physical principle to allocate them. We believe their role is to "connect" fermions in different blocks since the fermions within each fermion block have "known" interactions. Note that there are initially eight blocks, each with their own set of symmetry groups and corresponding interactions, and initially no interactions between the eight blocks. If there are no block interactions (except gravity), then the Physics of the fermion set is conceptually disjoint. Thus we choose to implement inter-block interactions as in Fig. 11.2 following the principle:

A fermion in any block has interactions either directly, or indirectly, with every other fermion in any other block.

Fig. 11.2 shows an implementation of this principle. The horizontal lines in Fig. 11.2 indicate 1:1 transformations between all corresponding fermions of each "Normal" and "Dark" block. The three "angled" lines indicate 1:1 transformations between corresponding fermions of a "Normal" and a "Dark" fermion block in the layer above it. The result is a see-saw pattern.

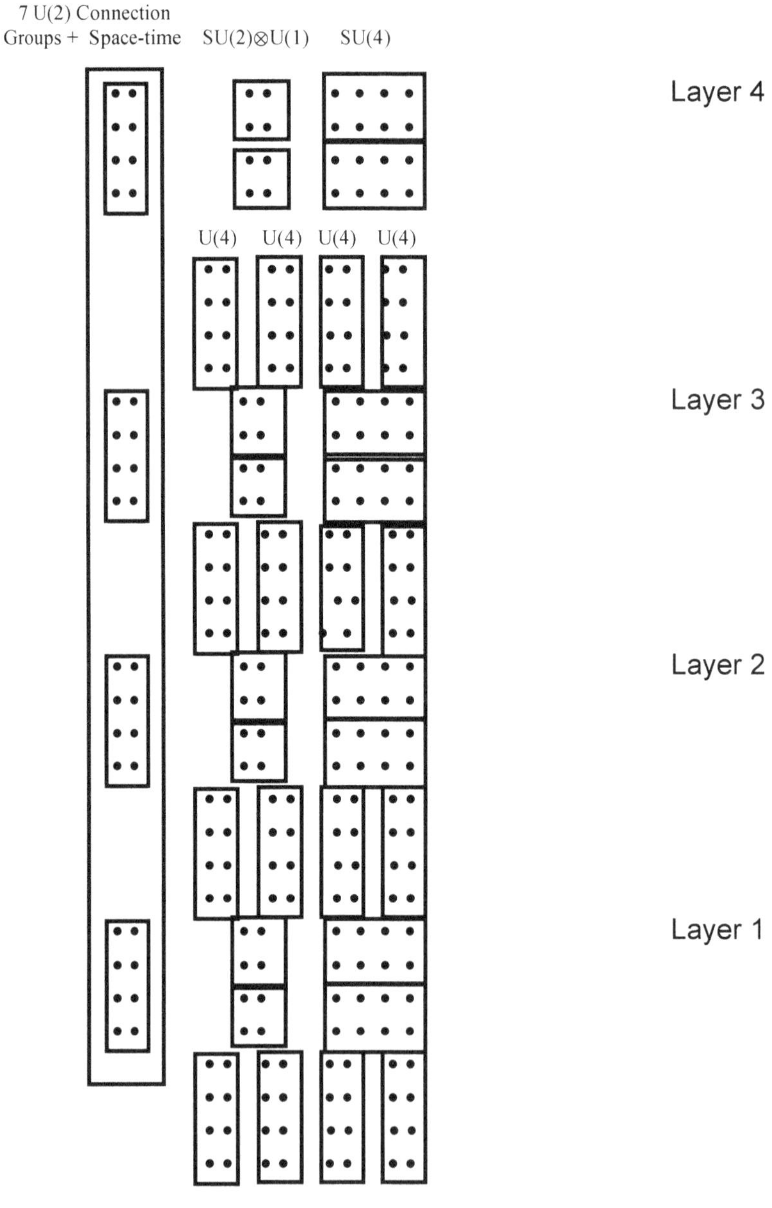

Figure 11.1. The four layers of QUeST, NEWUST and NEWQUeST internal symmetry groups (and space-time) with SU(4) before breakdown to SU(3)⊗U(1). Note the left column of blocks are combined below to specify a 4

dimension real space-time plus seven U(2) Connection groups. Note each layer has 64 dimensions = 56 + 8 dimensions.

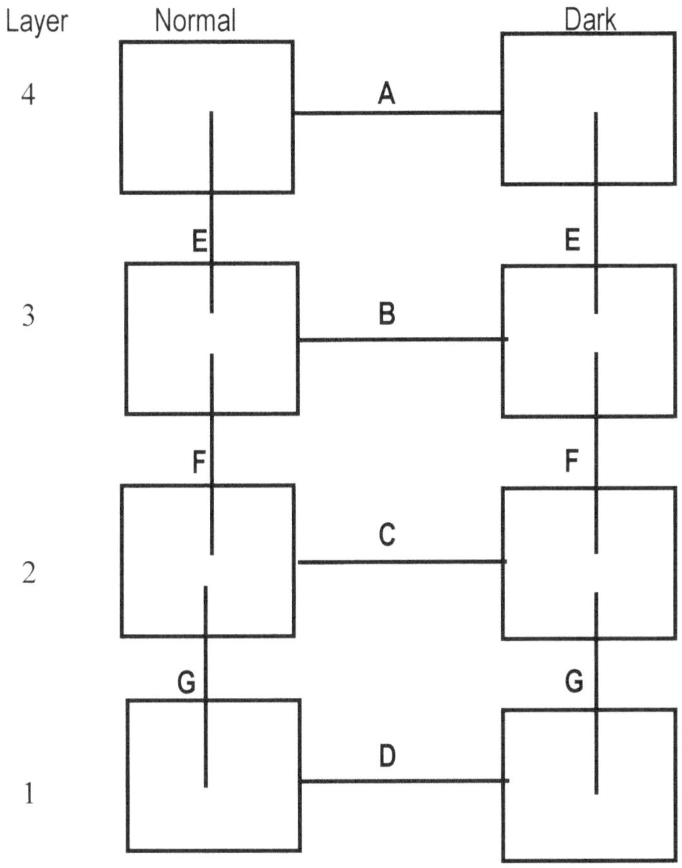

Figure 11.2. The seven U(2) Connection groups (shown as 10 lines) between the eight NEWQUeST/NEWUST blocks. Connection groups are obtained by transfering 28 dimensions from QUeST space-time to internal symmetries with the consequent reduction of the space-time from four octonion (complex quaternion) coordinates to four real coordinates. The Connection groups generate rotations and interactions between corresponding fermions and vector bosons of each pair of blocks. The Normal and Dark sector U(2) vertical connections above (E, F, G) represent the same U(2) groups.

11.1.1 The U(2) Connection Groups

The seven U(2) Connection groups of Fig. 11.2 generate "rotations" and interactions between *corresponding* fermions and vector bosons of each pair of blocks of the eight blocks of fermions in NEWQUeST/NEWUST.

11.1.1.1 Horizontal Lines

The horizontal lines in Fig. 11.2 (A, B, C, and D) each represent a U(2) Connection group that "rotates" two *corresponding* fermions in the Normal and Dark sectors of each layer. Thus a Normal e is "rotated" with a corresponding Dark e, and so on.

Each of the four horizontal Connection Groups has a reducible U(2) representation D that is the sum[58] of 4*8 = 32 irreducible U(2) representations. We may view the irreducible representations D_j as strung along the diagonal.

$$D = \sum_{j=1}^{32} D_j \tag{11.1}$$

for each of the U(2) groups of the four horizontal lines in Fig. 11.2.

The U(2) group also specifies gauge field interactions between corresponding fermions in each layer of the Normal and Dark sectors of the form

$$g\overline{\Psi}_{Nn}\gamma\cdot A\cdot T\Psi_{Dn} \tag{11.2}$$

where N indicates a Normal fermion and D indicates the corresponding Dark fermion, with A being a U(2) gauge vector boson, and n the label for corresponding fermions.

These U(2) transformations imply that the Normal and Dark sectors have the same species.

11.1.1.2 Vertical Lines

The pairs of vertical lines in Fig. 11.2 (E, F, G) each represent a U(2) Connection group that "rotates" sets of two *corresponding* fermions in adjacent layers as shown in Fig. 11.2 in the Normal and Dark sectors. Thus a Normal e in layer 1 is rotated with a corresponding Normal e in layer 2, and so on.

Each of the three (six counting both Normal and Dark lines in Fig. 11.2) vertical Connection Groups has a reducible U(2) representation D that is the sum of 64 irreducible U(2) representations.[59] We may view D as an array of 64 U(2) irreducible representation dimensions D_j strung along the diagonal.

$$D = \sum_{j=1}^{64} D_j \tag{11.3}$$

for each of the U(2) groups of the 3 (6) horizontal lines in Fig. 11.2. Note the 64 irreducible representations include both Normal and Dark sectors of a layer.[60]

Each U(2) group also specifies a gauge field interaction between corresponding fermions in adjacent layers for both Normal and Dark sectors:

---

[58] Eight fermions per generation • four generations, thus accounting for each fermion in a block.

[59] Eight fermions per generations • four generations • 2 types of matter (Normal and Dark).

[60] There are 64 fermions in total for each of the four layers of NEWQUeST/NEWEST.

$$g\overline{\Psi}_{nl_1}\gamma\cdot A\cdot T\Psi_{nl_2} \qquad (11.4)$$

where l_1 and l_2 designate layers, A is a gauge field vector boson, and n the label for corresponding fermions.

Each E, F, and G U(2) group reducible representation includes both Normal and Dark sectors.

11.1.2 The Connection Groups are UltraWeak Interactions

Since there is no convincing experimental evidence for particle interactions between Normal and Dark matter, or between Normal fermion layers the Connection groups appear to be Ultraweak.

4.2 NEWUTMOST with Six Real Space-Time Coordinates (Dimensions)

The NEWUTMOST space is a Blaha number N = 6 HyperCosmos space. It is the Megaverse (Multiverse) with 6 real-valued space-time coordinates by eq. 9.3 above. It has a d_{d6} = 1024 dimension array.

We view the NEWUTMOST dimension array as composed of four copies of the N = 7 dimension array. Initially 4·32 = 128 dimensions are for space-time coordinates.

We allocate 4·28 = 112 dimensions to each of the four NEWQUeST copies within NEWUTMOST to give each copy 7 SU(2) Connection groups. Therefore 4* 28 = 112 of the 128 space-time dimensions of UTMOST are mapped. The remaining 16 dimensions give 6 real-valued space-time dimensions to the Megaverse space, 8 dimensions to a Megaverse SU(4) Connection group, and 2 dimensions to a U(1) group for all fermions in the Megaverse. See Figs. 11.3 and 11.4.

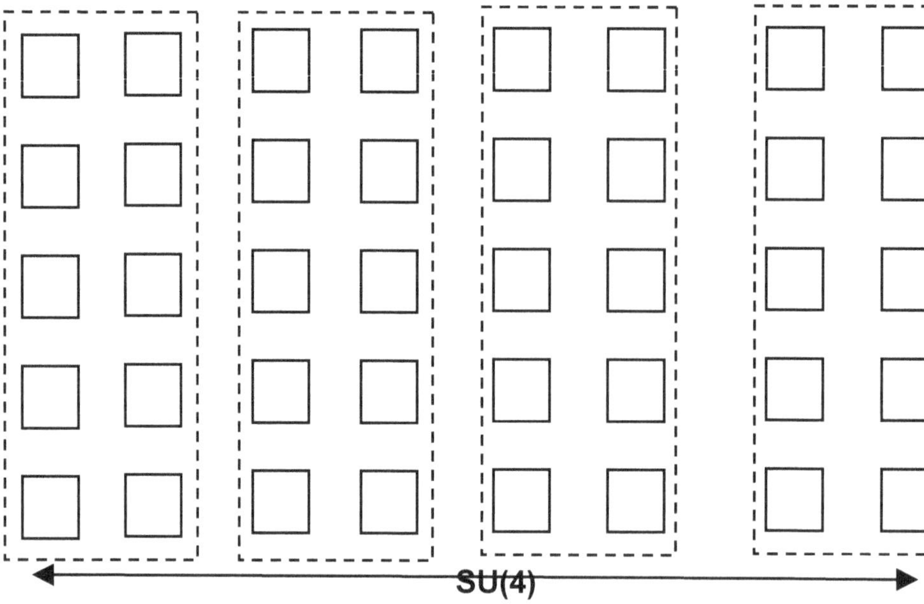

Figure 11.3. NEWUTMOST has four NEWQUeST copies. An SU(4) internal symmetry Connection group maps between corresponding fermions in the four copies: fermion by fermion. The U(1) Connection group applies to every fermion. It is not shown in this figure.

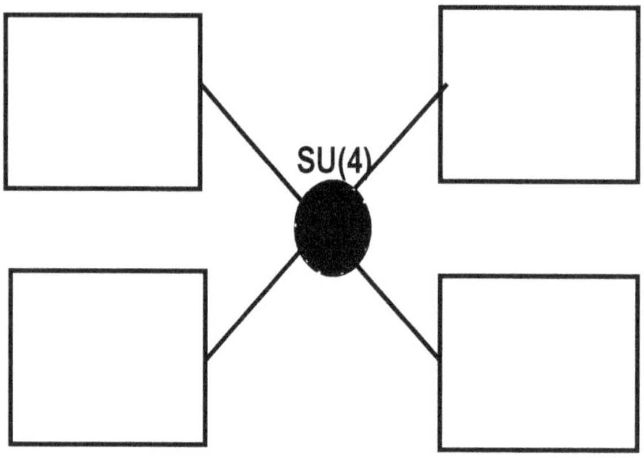

Figure 11.4. The SU(4) Connection Group of NEWUTMOST connecting fermions in the four NEWQUeST "copies" blocks.

4.3 NEWMaxiverse with Eight Real Coordinates (Dimensions)

The NEWMaxiverse is a Blaha number N = 5 space with a dimension array of d_{d5} = 4096 dimensions.

The NEWMaxiverse contains four copies of NEWUTMOST. Each NEWUTMOST has 4*6 = 24 space-time dimensions. (The remainder in each NEWUTMOST copy consists of internal symmetries and Connection groups.) We allocate the 24 dimensions to eight real space-time dimensions plus 16 dimensions for a new ultraweak (possibly broken) SU(8) Connection group for the four parts of the NEWMaxiverse. See Fig. 11.5.

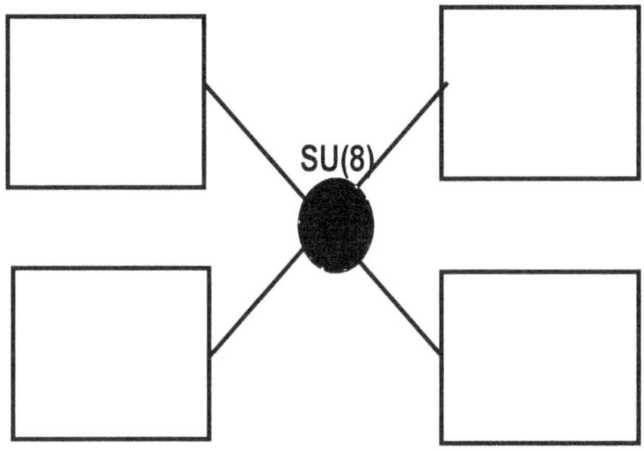

Figure 11.5. The SU(8) Connection Group of NEWMaxiverse connecting fermions in the four NEWUTMOST "copies" blocks.

4.4 Determining the Connection Groups for a Space

In the previous sections we determined the Connection Groups for N = 7, N = 6, and N = 5 spaces. The dimensions required for the Connection Groups of a space d_{gN} for $N \leq 6$ can be obtained by noting the space-time dimension of a space-time r is related to the space-time dimension r – 2 of the next lower space. Since a space is four copies of the next lower space it has 4(r – 2) space-time dimensions from the four duplicates. Then

$$4r_{N+1} = 4(r_N - 2) = r_N + d_{gN}$$

or

$$d_{gN} = 3r_N - 8 \qquad (11.5)$$

where d_{gN} is the space-time reallocated to Connection Groups in space N.

Eq. 11.5 determines the dimensions available for Connection Groups for the space with Blaha number $N \leq 6$. Using eq. 9.3:

$$r_N = \log_2 (d_{dN}/16) \qquad (9.3)$$

we relate d_{gN} to the size of the dimension array:

$$d_{gN} = 3 \log_2(d_{dN}/16) - 8 \qquad (11.6)$$

thus specifying the number of dimensions allocated to space number N.

12. Fractional Dimensions: Spaces and Particles

12.1 Enhanced HyperCosmos Spectrum

The HyperCosmos of Fig. 1.2, which we derived in Blaha (2022a), suffices for the description of Cosmology in the large. However, the extension of the spectrum to fractionally dimensioned arrays suggests a new, deeper layer supporting a new paradigm for the Cosmos and Physics. We will describe fractional array dimensions, spaces, and particles in this chapter. The chapter ends with the suggestion that physical examples may exist in Condensed Matter Physics in quantum spin liquids.

A major benefit of the HyperCosmos is its support for a deeper level based on the decomposition of particles, operators and dimensions into component parts down to the infinitesimal level. Thus one can view the generation of HyperCosmos spaces and universes (and particles) as a *local* phenomenon that grows to enormous proportions.

A possible application is parton-like models of Deep Inelastic scattering. Another apparent application is in Quantum Spin Liquids in the formal description of *spinons*.

Fractionation opens the door to the deeper investigation of the "interior" of fundamental elementary particles, and also to the *very* beginning of HyperCosmos universes. Blaha (2021d) describes initial stages of our universe where it appears to display free particle-like behavior. We see universes as expanding particles and particles as fixed size universes.

This chapter defines a fractionation method for dimensions, spaces, and particles. The method provides a procedure to "chop" these items into pieces of a specific fractionality. One can chop an item into halves, quarters, eights, and so on, up to infinite granulation.

This formulation differs from lattice gauge theories, which subdivide space-time, for the purpose of performing calculations in existing models in Quantum Field Theory. Fractionation subdivides physical "things" such as particles, spaces (including all symmetries) and dimensions.

Quite simply, one can view lattice theory as dividing space-time for the purpose of performing calculations in known theories using space-time lattices. Fractionation divides the interior of entities, such as particles, into parts, for deeper purposes.

12.2 Origin of Fractionality

Their origin is based on the observation that a partial sum of dimension array d_{cd} column (or row) elements (with non-negative space-time dimensions) as in Fig. 12.1 is always less than the number of columns in the space immediately above it by exactly 2 dimensions. For example the sum of the columns in column d_{cd} for Blaha numbers 9 through 5 is 62 while the next column size above it with Blaha number 4 is $d_{cd} = 64$. The complete set of column sizes d_{cd} for Blaha numbers 9 through 0 sums to 4094 while the next column size above it if the spectrum was increased to Blaha number N = -1 would be d_{cd} = 4096. If we embed the product of the 10 spaces in a SU(2048) fundamental representation then there is a shortfall of 2 real-valued dimensions.

We may remedy the shortfall by extending the spectrum infinitely to fractional dimension arrays "up to" N = ∞. The summation of the additional fractional array columns for all space-time dimensions less than zero equals 2 remedies the shortfall:

$$2 = 1 + ¼ + 1/8 + 1/16 + \ldots = 1 + 1$$

In general the d_{cd} column lengths satisfy the identity

$$\sum_{n=k+1}^{\infty} d_{cd}(n) = d_{cd}(k) \tag{12.1}$$

where n and k are Blaha space numbers.

The introduction of negative space-time dimensions requires further consideration, which we provide in chapter 13.

EXTENDED FORM OF THE HYPERCOSMOS SPACES SPECTRUM

	Blaha Space Number N = o_s	Cayley-Dickson Number n	Cayley Number d_c	Dimension Array column length d_{cd}	Space-time-Dimension r	CASe Group $su(2^{r/2},2^{r/2})$ CASe
	0	10	1024	2048	18	su(512,512)
	1	9	512	1024	16	su(256,256)
	2	8	256	512	14	su(128,128)
	3	7	128	256	12	su(64,64)
	4	6	64	128	10	su(32,32)
	5	5	32	64	8	su(16,16)
	6	4	16	32	6	su(8,8)
	7	**3**	**8**	**16**	**4**	**su(4,4)**
	8	2	4	8	2	su(2,2)
	9	1	2	4	0	su(1,1)
EXTENSION:						
	10	0	1	2	-2	U(1)
	11	-1	½	1	-4	u(½)
	12	-2	¼	½	-6	•
	13	-3	1/8	¼	-8	•
	14	-4	1/16	1/8	-10	•

•
•
•

Figure 12.1. The Extended HyperCosmos space spectrum.

12.3 The Partition of a Unit into Sequence of Fractional Space Units

In chapter 4 we described the enhanced HyperCosmos spectrum that adds an infinity of spaces to the HyperCosmos spectrum. Most of the additional spaces are fractional. Fig. 12.2 depicts the amalgamation of spaces of negative space-time dimension from Blaha number ∞ to Blaha number N = 12 to construct a unit space with $d_{cd} = 1$ corresponding to Blaha number N = 11.

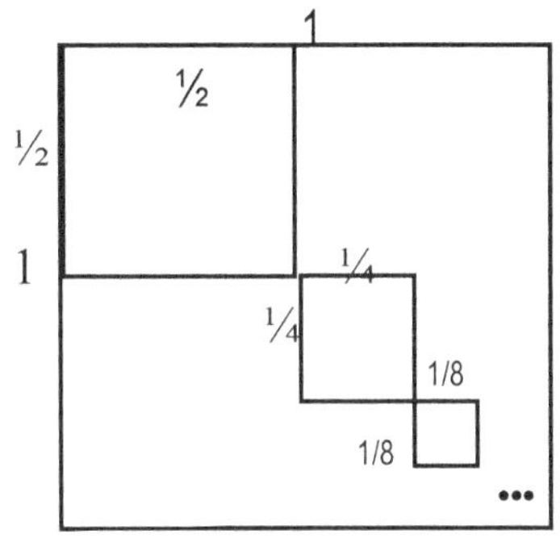

Figure 12.2. Depiction of a amalgamation of an infinite number of spaces from Blaha number N = ∞ through Blaha number N = 12 to form a space with a 1×1 dimension array.

The construction of a unit space follows from eq. 12.2:

$$\sum_{n=12}^{\infty} d_c(n) = d_c(11) = 1 \tag{12.2}$$

where n, 11 and 12 are Blaha space numbers.

We will use the expansion of the unit space to disassemble fundamental fermions and bosons, raising/lowering operators, and dimensions into Gold Dust.

12.4 Creation of a Dust Assemblage

One can create an assemblage of dust for a particle or dimension as a finite sum of items or spaces, or as an infinite sum.

12.4.1 A Finite sum of a Specific Fractional Level

A particle or space can be directly expressed as a finite sum from examination of the HyperCosmos spectrum of spaces in Fig. 12.1. For example $d_{cd}(3) = 256$ can be disassembled as four 64 element spaces:

$$d_{cd}(3) = d_{cd}(5) + d_{cd}(5) + d_{cd}(5) + d_{cd}(5)$$

where the Blaha number 3 space is fractionated into four number 5 spaces. Another example of finite disassembly is

$$d_{cd}(11) = d_{cd}(14) + d_{cd}(14) + d_{cd}(14) + d_{cd}(14) + d_{cd}(14) + d_{cd}(14) + d_{cd}(14) + d_{cd}(14)$$

where $d_{cd}(11) = 1 = 8\ d_{cd}(14) = 8 \cdot 1/8$.

12.4.2 An Infinite Assemblage

The creation of a "Gold Dust" infinite assemblage for a specific fractional level follows from repeated use of eq. 12.1. Consider one item of one of the sets of fundamental fermions and bosons, raising/lowering operators, and dimensions. Suppose it is of Blaha number N where .

$$d_{cd}(N) = 2^{11-N} \tag{12.3}$$
$$d_{cd}(j) = 2^{11-j}$$

Suppose we wish to fractionate it into $dust_{jN} = d_{cd}(j)d_{cd}(N)^{-1}$ parts of size $sdust_N = d_{cd}(N)$ where $j < N$. Note

$$d_{cd}(j) = dust_{jN} \cdot sdust_N \tag{12.4}$$

We first use eq. 12.1 to obtain

$$\sum_{n=N+1}^{\infty} d_{cd}(n) = d_{cd}(N) \tag{12.5}$$

$$\sum_{n=j+1}^{j\infty} d_{cd}(n) = d_{cd}(j) \tag{12.6}$$

implying

$$\sum_{n=j+1}^{N} d_{cd}(n) = d_{cd}(j) - d_{cd}(N) \tag{12.7}$$

where N, n and j are Blaha space numbers. Then

$$d_{cd}(j) = \sum_{n=j+1}^{N} d_{cd}(n) + d_{cd}(N) \tag{12.8}$$

expresses $d_{cd}(j)$ in terms of the more fractionated $d_{cd}(n)$ of higher Blaha number spaces. Each of the terms on the left side of eq. 12.8 represents a smaller space.

We now can fractionate each of these terms $d_{cd}(n)$ in the summation by iterating eq. 12.7 repeatedly. Thus $d_{cd}(j)$ is fractionated into a sum of $dust_{jN}$ parts consisting of $sdust_N$ dust for some Blaha space level N.

Clearly we can repeat this decomposition process over and over creating level after level of decomposition. Thus every particle creation operator, annihilation operator, and dimension, can be subdivided to dust of various levels without limit.

Every item can be viewed as an assemblage of dust including the possibly an infinity of infinitesimal dust "grains." A particle, which we can represent as one item with $d_{cd}(11) = 1$, can be disassembled into a countable infinity, ∞, of infinitely small dust, which we represent as $d_{cd}(\infty) = 1/\infty$, or disassembled into a large, but finite number, of dust grains.

The benefit of the fractionation process will become apparent later in section 12.8 when we construct wave functions for fractionated particles. We call fractionated particles, and dimensions, *gold dust*.

12.5 Types of Gold Dust

The fractionation process may be applied to individual or sets of creation/annihilation operators, fermions, bosons, and dimensions.

The types of the quantities subject to fractionalization are:

Particles
Creation operators of fundamental particles.
Annihilation operators of fundamental particles.
Internal Symmetry groups and their fundamental representation dimensions.
Dimensions

Gold dust of the first four types is quantum with commutation rules. Dimensions gold dust is not quantum.

The fractionation of a particle reduces a particle to an aggregate of "gold dust" where each infinitesimal part contains part of the particle's charge and internal symmetries.

Particle gold dust differs due to differences in the internal symmetries of particles. There is electron gold dust, neutrino gold dust, and gold dust of the various quark particle types: three up-quark types and three down-quark types as well as gold dust specific to the various fermion generations and layers.. There is also a variety of Dark gold dust.

12.6 Confinement of Gold Dust

A fundamental particle, as we currently see it, is a unitary entity that does not appear to be composite although there are theories that subdivide particles. Therefore it may have a form of confinement: *dust confinement*. Physicists are somewhat familiar with quark confinement, which may be a related phenomenon, although the confinement mechanisms for dust and quarks are problematic. Some like Kenneth Wilson (Cornell and Ohio State) suggest a lattice approach. Some, like the author, suggest it is dynamic through higher derivative strong interactions. (Wilson has suggested to the author that the author's theory is a phenomenology for his lattice theory. Perhaps.)

We now have a new confinement problem: How do particles, which are subject to fractionation as we suggested earlier, cohere as whole objects to become the fundamental particles of NEWQUeST (NEWUST) and NEWUTMOST?

A particle, composed of Gold Dust, presumably has total dust confinement. The dust may be confined by a force or by being enclosed in an impenetrable membrane or by confinement through some lattice theory.

Gold Dust confinement thus raises an important issue. Perhaps it is an issue that experiment has raised already but we were unable to recognize. Perhaps parton models of Deep Inelastic Scattering are about fractionated quarks. It is known that parton models have some issues when matched to experiment. Perhaps the quarks in a proton or neutron are being viewed at higher precision (higher fineness) revealing their internal fractional gold dust components. See section 6.3.

Gold dust "glue" may be analogous to color confinement "glue." Then confined Gold Dust may aggregate to form particles.

12.7 The Profile of a Particle as it Proceeds to Higher Fineness through Increased Fractionation

If we fractionate a fundamental particle we may expect the particle to pass through a number of stages if we could view it in a microscope of increasing resolution. First it becomes a set of quantum "sub-particles. Then it passes through the stages:

Quantum matter/liquid – with a lumpiness still
Classical matter/liquid – a smooth "classical" fluid due to the many parts that obscure the quantum aspect.

In the final states of fractionation the Gold Dust is so fine that it appears to be a classical fluid[61] thus accomplishing the reduction of a particle to a fluid of infinitesimal parts. The matter/fluid (dust) has spins, internal quantum numbers, species types, and energies distributed throughout. *Thus each type of fermion and boson has its own form of dust.*

Gold Dust combines to produce the creation/annihilation operators of particles. We may view the particles of a space as created in a Fundamental Frame in Blaha (2029) at the origin point of a universe. (Universes begin as a point composed of dust.) At the origin of a space (universe) we see a spontaneous condensation of dust to particles. Then splitting of the space into copies occurs as described in Blaha (2022) followed by dynamical evolution with interactions as the universe expands.

12.8 Wave Functions and Commutation Rules for Particle Gold Dust

At every level of fineness we can define fractional particle states and creation/annihilation operator commutation rules.

We assume a particle exists in a space of Blaha number N. We consider a particle and treat it as a one dimension space of Blaha number 11 with $d_{cd} = 1$. We can assume it can fractionate to d_{cd} (o) = 1/n for some even integer o using

$$d_{cd}(o) = 2^{11-N} = 1/n \tag{12.9}$$

---

[61] See eqs. 12.14a and 12.19a below that demonstrate classical field limits of quantum fields.

12.8.1 Fractional Boson Fields

For the moment we omit momenta. Then to create a boson particle of fraction 1/n from the vacuum state |0> we define a 1/n fraction state with

$$|1/n\rangle = a^{1/n\dagger}|0\rangle \tag{12.10}$$

A two-particle state has the form

$$|1/n_2, 1/n_1\rangle = a^{1/n_2\dagger}a^{1/n_1\dagger}|0\rangle \tag{12.11}$$

and we define an annihilation with

$$a^{1/n}\,|1/n\rangle = |0\rangle \tag{12.12}$$
$$a^{1/m}\,|0\rangle = 0 \tag{12.13}$$

With this basis one can define commutation (or anti-commutation) relations for fractional operators. We now introduce momenta, q and p in the space of Blaha number N with space-time dimension r. Then we have the commutation relations for boson creation/annihilation operators:

$$[a^{1/n}(q), a^{1/m\dagger}(p)] = \delta_{nm}\delta^{r-1}(q - p)/m \tag{12.14}$$

$$[a^{1/n}(q), a^{1/m}(p)] = 0$$

$$[a^{1/n\dagger}(q), a^{1/m\dagger}(p)] = 0$$

Note fractional operators of different fractionation commute. Note also that in the limit $m \to \infty$ all operators commute:

$$\lim_{m \to \infty} [a^{1/n}(q), a^{1/m\dagger}(p)] = 0 \tag{12.14a}$$

We now define

$$a(p) = \sum_{\substack{n=4 \\ n\ \text{even}}}^{\infty} a^{1/n}(p) \tag{12.15}$$

where n ranges over all even numbers ≥ 4 with the result

$$[a(p), a(q)^{\dagger}] = \sum_{\substack{n=4 \\ n\ \text{even}}}^{\infty} 1/n \;= \delta^{r-1}(q - p) \tag{12.16}$$

$$[a(p), a(q)\,] = 0$$
$$[a(p)^{\dagger}, a(q)^{\dagger}] = 0$$

thus recovering the conventional boson commutation relations.

We next define fractional PseudoQuantum free boson fields in the space of Blaha number N:

$$\varphi_1{}^{1/n}(x) = \Sigma_\alpha\, [a^{1/n}{}_{1\alpha}\, f_\alpha(x) + a^{1/n}{}_{1\alpha}{}^\dagger f_\alpha{}^*(x)] \qquad (12.17)$$
$$\varphi_2{}^{1/n}(x) = \Sigma_\alpha\, [a^{1/n}{}_{2\alpha} f_\alpha(x) + a^{1/n}{}_{2\alpha}{}^\dagger\, f_\alpha{}^*(x)]$$

where the $f_\alpha(x)$ are Fourier fields with α being the momentum in Blaha space N.

Note that eq. 12.14a implies that

$$\lim_{n \to \infty} \varphi_i{}^{1/nn}(x) = \text{a classical field} \qquad (12.17a)$$

for i = 1, 2. Thus the limits of boson fields are classical fields.

12.8.2 Fractional Fermion Fields

We may follow a similar procedure for a fermion field. in a space of Blaha number N. We consider a fermion particle and treat it as being a one dimension space of Blaha number 11 with $d_{cd} = 1$. Assuming fractionation to d_{cd} (o) = 1/n for some even integer n we develop anti-commutation relations:

$$\{b^{1/n}(q), b^{1/m\dagger}(p)\} = \delta_{nm}\delta^{r-1}(q - p)/m \qquad (12.18)$$

$$\{b^{1/n}(q), b^{1/m}(p)\} = 0$$

$$\{b^{1/n\dagger}(q), b^{1/m\dagger}(p)\} = 0$$

and similarly for $d^{1/n}$ and $d^{1/n\dagger}$:

$$\{d^{1/n}(q), d^{1/m\dagger}(p)\} = \delta_{nm}\delta^{r-1}(q - p)/m \qquad (12.19)$$
$$\{d^{1/n}(q), d^{1/m}(p)\} = 0$$
$$\{d^{1/n\dagger}(q), d^{1/m\dagger}(p)\} = 0$$

Note fractional operators of different fractionation anti-commute. Note also that in the limit m → ∞ all operators commute:

$$\lim_{m \to \infty} \{d^{1/n}(q), d^{1/m\dagger}(p)\} = \lim_{m \to \infty} \{b^{1/n}(q), b^{1/m\dagger}(p)\} = 0 \qquad (12.19a)$$

We now define

$$b(p) = \sum_{\substack{n=4 \\ n\ \text{even}}}^{\infty} b^{1/n}(p) \qquad (12.20)$$

where n ranges over all even numbers ≥ 4 with the result

$$\{b(p), b(q)^\dagger\} = \sum_{\substack{n=4 \\ n \text{ even}}}^{\infty} 1/n = \delta^{r-1}(q-p) \tag{12.21}$$

$$\{b(p), b(q)\} = 0$$

$$\{b(p)^\dagger, b(q)^\dagger\} = 0$$

thus recovering the conventional fermion commutation relations.

We next define fractional PseudoQuantum free fermion fields in the space of Blaha number N:

$$\psi_1^{1/n}(x) = \Sigma_\alpha\, [b^{1/n}{}_{1\alpha}\, f_\alpha(x) + d^{1/n}{}_{1\alpha}{}^\dagger f_\alpha{}^*(x)] \tag{12.22}$$
$$\psi_2^{1/n}(x) = \Sigma_\alpha\, [b^{1/n}{}_{2\alpha} f_\alpha(x) + d^{1/n}{}_{2\alpha}{}^\dagger\, f_\alpha{}^*(x)]$$

where the $f_\alpha(x)$ are Fourier fields with α being the momentum in Blaha space N.

Note that eq. 12.19a implies that

$$\lim_{n\to\infty} \psi_i^{1/n}(x) = \text{a classical field} \tag{12.22a}$$

for i = 1, 2. Thus the limits of fermion fields (and similarly of boson fields) are classical fields. As a result calculations of fermion and boson dynamics in the dust within particles are classical. We found a similar result for universe particles at their creation point in Blaha (2021d).

Thus we have a basis for a fractional creation/annihilation operator formalism for free fermions and bosons in a space-time of dimension r.

12.8.1 Possible Dust Confining Interaction Within a Particle

We have suggested that particles are composed of extremely fine dust. The possibility of a confining interaction that holds particles together then seems likely. We know that such an interaction must be very strong. From earlier work above we see that multiplets of leptons and quarks may have an SU(4) symmetry before symmetry breaking. In earlier work we considered a particle model of universes that exhibits universe expansion.[62] In that book we found the below content:

The universe vector interaction which we denote with the quantum field label Y with $e_U = (4\pi\alpha_U)^{1/2}$, and the other coupling constants for QED, ElectroWeak SU(2) and Strong SU(3) have a remarkable regularity—they double from interaction to interaction as Fig. 7.1 shows. It is interesting that Cayley numbers in Octonion Cosmology also display doubling. The relation between these doubling phenomena, if any, merits further investigation..

[62] Blaha (2021d).

INTERACTION	COUPLING CONSTANT[63]
Y Interaction e_U	0.152
QED $e_{QED} = (4\pi\alpha_{QED})^{1/2}$	0.303
Weak SU(2) g_W	0.619
Strong SU(3) g_S	1. 22

Figure 7.1. The interaction constants show a regular doubling. The cause of the doubling is not apparent.

where we propose a coupling constant for universe expansion: Y Interaction e_U.

Given the evident doubling see above we suggest a new entry for the proposed SU4) confining interaction of dust in particles:

INTERACTION	COUPLING CONSTANT[64]	
Ultra Strong SU(4) g_{US}	2.44	(12.23)

Based on the previous studies of quark confinement it seems reasonable to have an SU(4) group with an inter-dust gauge field interaction.[65]

Calculations of dynamics within particles may only require *classical* field theory in view of eq. 12.22a (and a similar result for boson fields). Thus we appear to have a tractable dynamics for intra-particle dynamics based on fractional fields.

12.9 CASe Groups: SU(n), su(n, n), U(1/n)

The groups associated with the HyperCosmos all have group numbers n or 1/n that are powers of 2. Thus they all support fractionation.

12.9.1 Fractionating CASe Groups: su(n, n)

The CASe groups that appear in Fig. 12.1 of the form of su(n, n) have 4n real-valued coordinates in their fundamental irreducible representations. It is important to note that n is necessarily an even number due to its relation to Cayley numbers. Because n is always even, and a power of 2, in HyperCosmos CASe groups, we can fractionate su(n,n) to su(n/2,n/2) and thence down to su(1,1) Then we can further fractionate it into the fractional groups U(1/n) in Fig. 12.1 that we discuss next.

The basis of the fractionation is the number of coordinates in the fundamental irreducible representation of a CASe group is a power of 2. Fractionation yields a CASe group with *half the number of coordinates* at each stage of fractionation. Thus the dimensions of the group representation is fractionated down to the su(1, 1) fraction.

The su(1, 1) representation has four real-valued coorinates. The reduction of su(1,1) therefore produces the Blaha number 10 U(1) CASe group as Fig. 12.1 illustrates. In the next section we describe fractionation for CASe U(1/n) groups.

12.9.2 Fractionating CASe Groups: U(1/n)

The U(1) group elements may be represented as:

---

[63] M. Tanabashi *et al* (Particle Data Group), Phys. Rev. D**98**, 030001 (2018).
[64] M. Tanabashi *et al* (Particle Data Group), Phys. Rev. D**98**, 030001 (2018).
[65] This SU(4) differs from the proposed SU(4) uniting lepton and quark multiplets.

$$e^{i\theta}$$

where θ ranges from 0 to 2π. Reducing the U(1) representation to U(½) requires a reduction in the range of coordinates. This process is accomplished by modifying the representation to

$$e^{i\theta/2}$$

The θ range remains 0 through 2π. Thus the resulting operators are factored by 2.

A similar procedure leads to the representation of U(1/n) where n is an even integer:

$$e^{i\theta/n}$$

Thus the CASe groups U(1/n) support fractionation of the coordinates of their group irreducible representations.[66]

12.9.3 Fractionating Symmetry Groups: SU(n)

HyperCosmos spaces have a dimension array that can be viewed as the fundamental representation of SU(n) where $n = d_{dn}/2 = 2^{21-2N}$ for some N. The dimension array group is separated into sets of symmetry groups in each HyperCosmos space. The symmetry groups are copies of color SU(4) assuming lepton-quark doublets before symmetry breaking and into SU(2)⊗U(1) ElectroWeak groups.

If we consider SU(n) in general we note that it can be fractionated indefinitely if n is an even number. U(1) also can be fractionated indefinitely. Thus the groups of interest in the set of symmetry groups of each space (universe) are exactly the ones susceptible to fractionation based on powers of 2^n.

SU(n) for odd n does not support indefinite 2^n fractionation with the sole exception being U(1). For odd n the fundamental representation coordinates after the first fractionation become fractions n/m that are not equal to powers of 2. Thus it would not be consistent with the 2^n fractionation of the HyperCosmos and thus not support a fractional dynamics based on 2^n fractionation.

12.10 A New Level for the HyperCosmos

The discovery of the basis of HyperCosmos features in Gold Dust implies a new foundation for Cosmology in an infinitesimally grained substratum. This substratum is very credible from a Philosophic viewpoint since it avoids instant creation in the large by *fiat*. The Cosmos grows infinitesimally—space by space, dimension by dimension, fractionally by creation operators, fractionally by particles, and fractionally in interactions. Annihilation takes place infinitesimally. All this with zero elapsed time since time does not exist at this level. The Cosmos is generated *locally in a manner appealing to the continuing local trend of Physics.*

---

[66] One could extend this discussion to U(1/a) where a is a real number.

The infinitesimal nature of creation and annihilation is masked by forms of aggregated confinement enforced by dynamics.

Gold Dust opens the issues of the binding of particles and the nature of negative space-time dimensions. Importantly, it raises the question of the confinement of Gold Dust within spaces, dimensions, and, most importantly, in fermion and boson particles.

The central role of creation and annihilation operators also leads to the consideration of such operators for the creation and annihilation of spaces (universes), which we considered in Blaha (2018e).

Our development of the HyperCosmos leads to a view of the Cosmos as composed of spaces (universes) that are dust, of dimensions that are dust, of universes of galaxies, clouds, stars and planets that are dust, of matter and energy that are dust, and of interactions that are dust. All things are Gold Dust united according to confinement and dynamics at multiple levels.

Looking backwards from the spaces (universes) of the HyperCosmos we see that their ultimate origin in Gold Dust provides a graduated platform of growth from "Nothingness."

12.11 Fractional Formalism for Quantum Spin Liquids?

Recently some studies[67] have suggested that cerium zirconium pyrochlore is a 3D quantum spin liquid with fractionalized spin excitations. The fractionalization apparently causes a particle such as an electron to split into two halves: quasiparticles called spinons that move throughout a crystal lattice. There appears to be dipolar and octopolar spinons with two "poles" and eight "poles" respectively.

Based on our fractional formalism it seems reasonable to consider it as providing a spinon formulation. Note that we have ½ type fractionation and 1/8 type fractionation. It would be interesting to see if spinons also have ¼ fractionation.

[67] Anish Bhardwaj *et al*, in npj Quantum Materials **7**, article 51 (2022).

13. Negative Space-time Dimensions

The concept of negative space-time dimensions in elementary particle physics is new. We define them in terms of a conventional operational definition of dimension. They enable the negative space-time dimensions of the FULL HyperCosmos spectrum to be understood.

The dimension of a space-time is the number of independent parameters needed to define a point. An over-determined space-time has a negative dimension.

In particular, we regard a negative space-time dimension as the equivalent of an over-specified point. Some negative space-time dimensions for r = -3 through r = 1 are.

r = -3

x = 3 x = 2*1.5 x = 6/2 x = 12/4 over-determined

r = -2

x = 3 x = 2*1.5 x = 6/2 over-determined

r = -1

x = 3 x = 2*1.5 over-determined

r = 0

x = 3

r = 1 One dimension space with one free parameter

x free parameter

r = 2 Two dimension space with two free parameters

x, y free parameters

where r is the space dimension.

12.1 Other Definitions of Dimension

Other definitions are possible. There are many. We believe the definition above is best suited for the HyperCosmos because it is in accord with the specification of dimension arrays. The 2 × 2, 1 × 1, and fractional array sizes indicate over-determined spaces.

14. Anti-Spaces and Anti-Universes

In chapter 2 we derived an energy spectrum for a Hydrogen-like atom. We mapped its positive energy states to the 10 HyperCosmos spaces. In chapter 12 we showed that the infinite energy states for Blaha number N > 9 map to a set of fractional spaces that have a physical interpretation.

We now address the set of negative energy states that we suggest map to anti-spaces and anti-universes. These states correspond to dimension arrays with a negative overall factor, $-d_{dN}$. They also have a set of HyperCosmos spaces corresponding to the spaces in Fig. 1.1.

Treating the dimension array elements as corresponding to coordinates of symmetry groups within a space we not interpret the negative sign as applying to fundamental symmetry group representation coordinates. Thus each coordinate is minus the corresponding coordinate in the "positive energy" analogue group representation.

We thus have

$$X_{negativeSpaceVector} = -X_{positiveSpaceVector} \tag{14.1}$$

This induces a trivial redefinition of group operators:

$$Operator_{negativeSpace} = -Operator_{PositiveSpace} \tag{14.2}$$

Since group multiplication is invariant under taking the negatives of all operators of the group, the groups of a negative space analogue are the same as those of the positive space. We will call the spaces corresponding to $-d_{dN}$ *anti-spaces*.

As a result we see:

1. The definition of anti-spaces starts with the same size dimension arrays, and the same space characteristics, as the 10 spaces in Fig.1.1.
2. The groups of an anti-space are the same as those of its space analogue.
3. An anti-space appears to be identical to its analogue prior to the onset of dynamics.

Anti-universes are defined as anti-space instances. Thus universes are *almost* the same as their analogue prior to the onset of dynamics. Universes have a positive charge (non-electric) as shown below. Anti-universes have a negative charge. Universes and anti-universes may interact through the equivalent of photons in a quantum electrodynamics of universes is defined. See Chapter 3.

14.1 Fermion Space Wave Functions

Given the identical characteristics of spaces and their corresponding anti-spaces we can define a fermion space wave function with

$$\psi_{iN_1N_2\alpha M}(\mathbf{x}, t, \mathbf{u}, t_u) = \Sigma_s \int d^{r1\text{-}1}k \int d^{r2\text{-}1}q\ \mathfrak{N}(q, k)[u_{r1}(k, s)\ b_{iN_1N_2\alpha M}(k, q, s)e^{-i(k\cdot x + q\cdot u)} + \\ + v_{r1}(k, s)d_{iN_1N_2\alpha M}(k, q, s)^{\dagger}e^{i(k\cdot x + +q\cdot u)}] \quad (14.3)$$

for i = 1, 2 where $\mathfrak{N}$ is a normalization constant, where m is the host fermion mass in the N_1 space's space-time and M is the mass-energy of the universe space of Blaha number N_2. Eq. 14.3 is identical to eq. 6.5 since the space and anti-space have the same characteristics.

The operator $d_{iN_1N_2\alpha M}(k, q, s)$ $(k, q, s)^{\dagger}$ creates an anti-universe and $b_{iN_1N_2\alpha M}(k, q, s)^{\dagger}$ creates a universe. Their symmetry groups are the same prior to dynamics.

There is a "charge" operator that differentiates between universes and anti-universes with a universe having charge 1 and an anti-universe having charge –1.

$$Q = \Sigma\ [b_{iN_1N_2\alpha M}(k, q, s)^{\dagger}b_{iN_1N_2\alpha M}(k, q, s) - d_{iN_1N_2\alpha M}(k, q, s)^{\dagger}\ d_{iN_1N_2\alpha M}(k, q, s)] \quad (14.4)$$

A universe – anti-universe pair is analogous to an electron-positron pair. One can define a universe dynamics similar to quantum electrodynamics with a universe interaction analogous to electromagnetism providing interactions and universe-anti-universe annihilation into photon-like particles.

The similarity to electrodynamics indicates that universes and anti-universes differ only in their charge.

14.2 Boson Space Wave Functions

Again given the identical characteristics of spaces and their corresponding anti-spaces we can define a "charged" boson space wave function with

$$\varphi_{iN_1N_2\alpha\beta M'}(\mathbf{x}, t, \mathbf{u}, t_u) = \int d^{r1\text{-}1}k \int d^{r2\text{-}1}q\ \mathfrak{N}(q, k)[a_{iN_1N_2\alpha\beta M'}{}^{+}(k, q)e^{-i(k\cdot x + q\cdot u)} + \\ + a_{iN_1N_2\alpha\beta M'}{}^{-}(k, q)^{\dagger}e^{i(k\cdot x + q\cdot u)}] \quad (14.5)$$

where + and – specify the charge and for i = 1, 2 where $\mathfrak{N}(q, k)$ is a normalization constant, where m′ is the host boson mass in the N_1 space's space-time and M′ is the mass-energy of the universe space of Blaha number N_2. Eq. 14.5 is similar to eq. 6.48 since the space and anti-space have the same characteristics.

The operator $a_{iN_1N_2\alpha\beta M'}{}^{-}(k, q)^{\dagger}$ creates an anti-universe and $a_{iN_1N_2\alpha\beta M'}{}^{+}(k, q)^{\dagger}$ creates a universe. Their symmetry groups are the same prior to dynamics.

There is a boson "charge" operator that differentiates between universes and anti-universes with a universe having charge 1 and an anti-universe having charge –1.

$$Q = \Sigma\ [a_{iN_1N_2\alpha M}{}^{+}(k, q, s)^{\dagger}a_{iN_1N_2\alpha M}{}^{+}(k, q, s) - a_{iN_1N_2\alpha M}{}^{-}(k, q, s)^{\dagger}a_{iN_1N_2\alpha M}{}^{-}(k, q, s)] \quad (14.6)$$

The similarity to the electrodynamics of charged bosons again indicates that universes and anti-universes differ only in their charge.

15. Splitting of Symmetries in HyperCosmos Universes

This chapter shows the origin of symmetry splitting is due to the chain of fermion-antifermion annihilations, to the origin of particles and symmetries in a Fundamental Reference Frame, and, particularly, is due to the structure of the dimension arrays derived from the structure of creation/annihilation operators in fermion wave function expansions. We developed this approach in Blaha (2021b).

15.1. Origin of Symmetries and the Fermion Spectrum

A HyperCosmos space has a dimension array with $d_{dN} = 2^{22-2N}$ elements. We can initially assume that it forms a fundamental representation of $SU(2^{21-2N})$. This group is split into a set of symmetry groups within a space. There are several points, upon which the symmetry splitting is based:

1. The set of HyperCosmos spaces has a combined dimension array (prior to splitting), which can be viewed as forming a fundamental representation of SU(2048) from Fig. 1.2. The spaces form an (infinite) direct product due to the splitting of SU(2048).

$$SU(2048) \rightarrow \prod_N SU(2^{21-2N})$$

   The ProtoCosmos uses a two dimension Hydrogen-like atom to indicate that the splitting may be due to a breakdown of scale invariance. The Nth space has a 2^{22-2N} element array of dimensions and thus may be viewed as a fundamental representation of $SU(2^{21-2N})$. The symmetries of this group is broken into various symmetry subgroups labeled by α:

$$SU(2^{21-2N}) \rightarrow \prod_\alpha SU(\alpha)$$

2. If one assumes the dimension arrays and groups were generated by a transformation from a Fundamental Reference Frame then the groups formed from dimension array coordinates are 2^{11-N} multiples of the Fundamental Reference Frame dimensions, which also number 2^{11-N} physical dimensions. (See section 9.5.) The dimension arrays total to $2^{11-N} \cdot 2^{11-N} = 2^{22-2N}$ elements in Blaha space N. Thus each dimension array may be viewed as split into 2^{11-N} blocks with copies of the Fundamental Representation fermions and symmetry groups respectively.

$$SU(2^{21-2N}) \leftarrow [G_{FundamentalReferenceFrame}]^{copies}$$

   where copies = 2^{11-N}.

Each copy of the symmetry groups is associated with each copy of the fundamental fermions in a 1:1 manner. Fermions in one copy do not use the groups of another copy for interactions. Each copy of the symmetry groups has an independent set of symmetries.[68] See Chapter 9 for a detailed example for our universe (N = 7).

3. The groups of a space are split in a manner reflecting the structure of the set of fermion creation/annihilation operators associated with the space. See chapter 4. Thus each set of b, $b^\dagger$, d, $d^\dagger$ corresponds to a splitting in 4's such as (e, up-quark triplet) and SU(2)⊗U(1).[69]

   Each set of b_1, $b_1^\dagger$, d_1, $d_1^\dagger$, b_2, $b_2^\dagger$, d_2, $d_2^\dagger$, due to PseudoQuantum wave functions, corresponds to a spitting in 8's such as (e, up-quark triplet and ν, down-quark triplet) and (SU(4) or SU(2)⊗U(1)⊗SL(1,C)).

   Taking account of fermion spin, each set of b_{1j}, $b_{1j}^\dagger$, d_{1j}, $d_{1j}^\dagger$, b_{2j}, $b_{2j}^\dagger$, d_{2j}, $d_{2j}^\dagger$ for j = spin up and spin down (r = 4) leads to a splitting into 16's such as (e, up-quark triplet and ν, down-quark triplet for Normal and Dark fermions) and (SU(4)⊗SU(2)⊗U(1)⊗SL(1,C)) separately for Normal and Dark sectors.

   For the case of r = 6 we find j takes on four values and there is a splitting into 32's leading to 4 generations of fermions (each containing e, up-quark triplet and ν, down-quark triplet) and the group product (SU(4)⊗SU(2)⊗U(1)⊗SL(1,C)⊗U(4)⊗U(4) taking account of the U(4) Generation and U(4) Layer groups. The fermion spectrum for our universe (Blaha number N = 7) in Figs. 15.1 – 15.3 show the separation in 4's, 8's, 16's, 32's and 64's also.

4. The relation between spaces—their quadrupling from space to space viewing their relation from a bottom-up perspective. This is evident in Fig. 1.2. See Fig. 15.4 for the quadruple pattern within a dimension array.

15.2 Bosons and Higgs Particles – Implementing Splittings?

The pattern of fermions and the pattern of symmetries show a remarkable similarity. Since scalar bosons appear to be necessary for symmetry breaking, and since the splittings that we have found provide limits to the range of interactions and appear to reflect mass differences in fermions and bosons, it is reasonable to have a set of Higgs bosons that equal the number of elements in the dimension array of space N: 2^{22-2N}. This number is the sum of the dimensions of the fundamental representations of the set of groups of the space.

For the HyperCosmos as a whole, we must take the total number of dimensions: 4096 real-valued dimensions in SU(2048). The first breakdown is to the HyperCosmos spaces in Fig. 1.2. Then within each space one encounters breakdowns for each broken

---

[68] This scenario is modified by the introduction of the Connection groups described earlier in chapter 11.

[69] The separation into 4's also reflects the separation due to space-time discussed in chapter 10.

symmetry group that emerges from the splittings. The splittings themselves are also the result of symmetry breaking.

15. 3 Evidence for the HyperCosmos

The patterns in the spectrum of fermions and symmetries, if they still hold when new particles and symmetries are found in higher generations and layers of our universe, will constitute indirect support for the UST, QUeST, and the HyperCosmos. See Chapter 1.

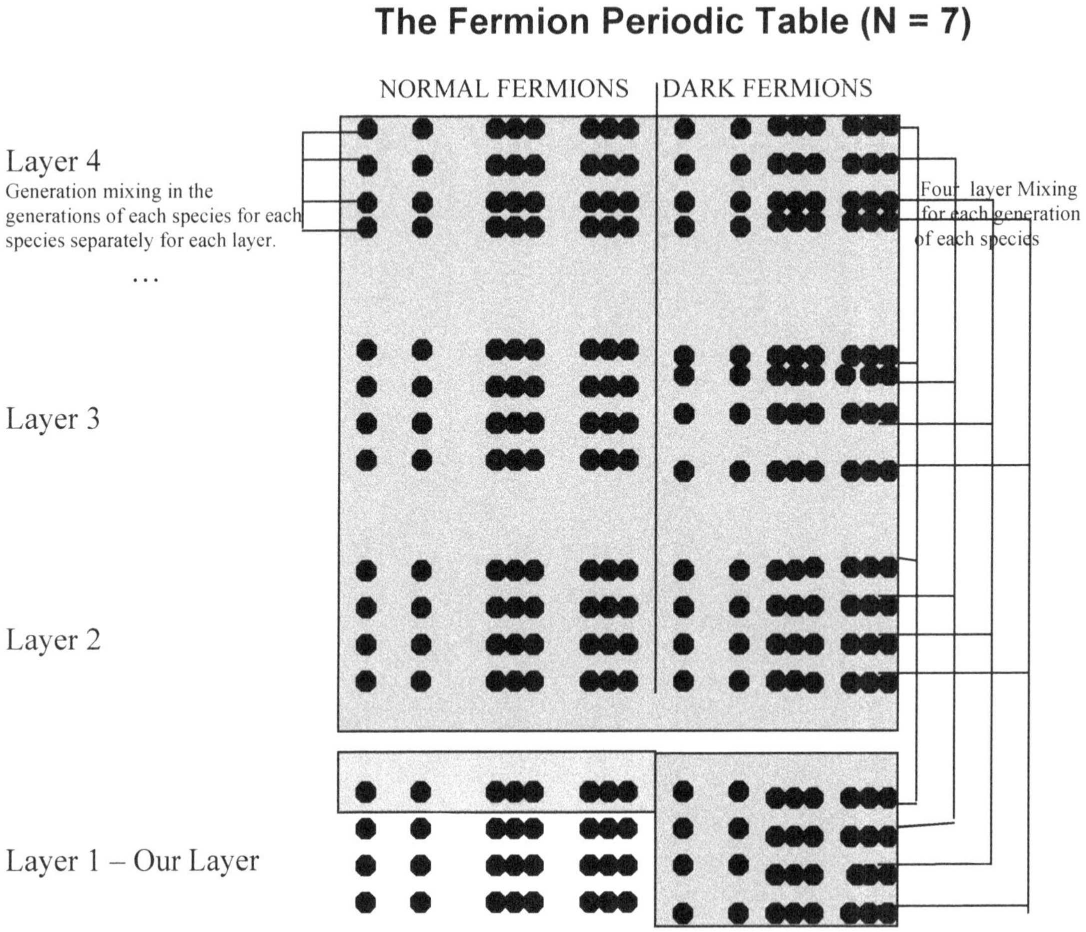

Figure 15.1. Fermion particle spectrum and partial examples of the pattern of mass mixing of the Generation groups and of the Layer groups. Unshaded parts are the known fermions, A shaded generation is an additional, as yet not found, 4th generation. The lines on the left side (only shown for one layer) display the Generation mixing within each layer. The Generation mixing occurs within each layer using a separate Generation group for each layer. The lines on the right side show Layer group mixing (for Dark matter) with the mixing among all four layers for each of the four generations individually. There are four Layer groups for Normal matter and four Layer groups for Dark matter.. There are 256 fundamental fermions. QUeST and UST have the same fermion spectrum. Each row is a 16 fermion subblock.

SU(4)- Based Fermion Particle Periodic Table

Number of Columns =	4 NORMAL	4	4 DARK	4
Layer 1 4	e 3 up-quarks	ν 3 down-quarks	e 3 up-quarks	ν 3 down-quarks
Layer 2 4	e 3 up-quarks	ν 3 down-quarks	e 3 up-quarks	ν 3 down-quarks
Layer 3 4	e 3 up-quarks	ν 3 down-quarks	e 3 up-quarks	ν 3 down-quarks
Layer 4 4	e 3 up-quarks	ν 3 down-quarks	e 3 up-quarks	ν 3 down-quarks

Figure 15.2. The fermion spectrum of our QUeST (and UST) universe. It corresponds directly with SU(4) (or SU(3)⊗U(1)) fermions. There are four layers. Each set of 4 fermions has 4 generations matching the number of rows in each layer. This Periodic Table is broken into Normal and Dark sectors.

NEWQUeST/NEWUST Symmetries Table

Number of Columns =	4 NORMAL	4	4 DARK	4
Layer 1 4	SU(4)⊗ Generation U(4)	SU(2)⊗U(1)⊗SL(2,C)[70] Layer U(4)	SU(4)⊗ Generation U(4)	SU(2)⊗U(1)⊗SL(2,C) Layer U(4)
Layer 2 4	SU(4)⊗ Generation U(4)	SU(2)⊗U(1)⊗SL(2,C) Layer U(4)	SU(4)⊗ Generation U(4)	SU(2)⊗U(1)⊗SL(2,C) Layer U(4)
Layer 3 4	SU(4)⊗ Generation U(4)	SU(2)⊗U(1)⊗SL(2,C) Layer U(4)	SU(4)⊗ Generation U(4)	SU(2)⊗U(1)⊗SL(2,C) Layer U(4)
Layer 4 4	SU(4)⊗ Generation U(4)	SU(2)⊗U(1)⊗SL(2,C) Layer U(4)	SU(4)⊗ Generation U(4)	SU(2)⊗U(1)⊗SL(2,C) Layer U(4)

Figure 15.3. The pattern of symmetries of NEWQUeST/NEWUST showing the splitting into 4's, 8's, 16's and 32's. Note each layer has its own set of internal symmetries. Normal and Dark sector symmetries are also different. The Layer Groups and the Connection Groups tie the various parts together. As shown, the pattern of symmetries closely matches the NEWQUeST dimension array splittings.The 4 ×4 blocks match the symmetry splits SU(4) appears split into SU(3)⊗U(1) experimentally.The eight SL(2, **C**) groups are transformed into one 4-dimension space-time and seven U(2) connection groups.

[70] The Lorentz group. $SO^{+}(1,3)$ in particular

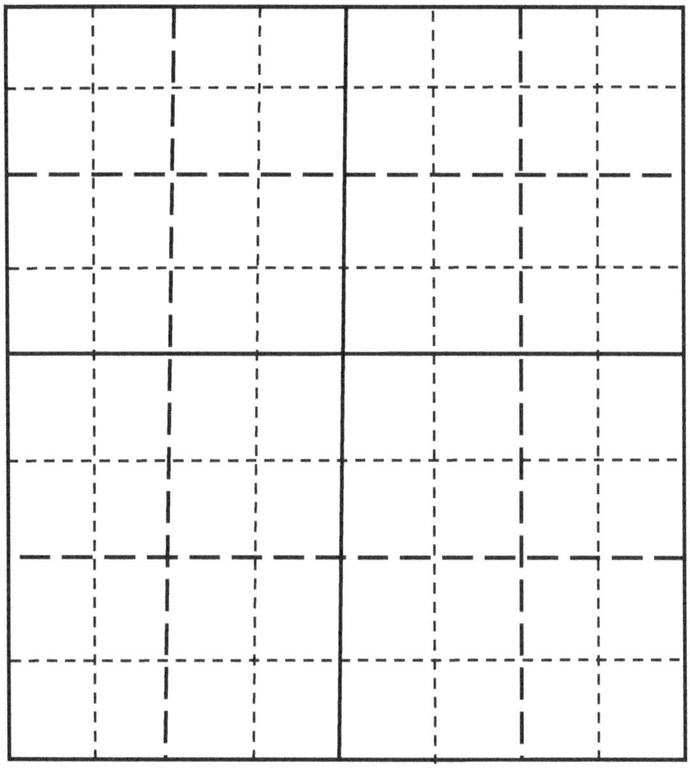

Figure 15.4. The Blaha number N = 6 dimension array quadrisected three times to N = 9 dimension arrays. Quadrisection continues down to N = 9. There are 64 N = 9 space parts visible.

16. QUeST – NEWQUeST – Unified SuperStandard Theory (UST)

The UnifiedSuperStandard Theory (UST) was created in the period before 2018 and described in books such as Blaha (2016b), (2018e), and (2020e). It had a large fermion spectrum and a large set of internal symmetries including the Generations groups and the Layer groups. The fermion spectrum consisted of four generations of fermions with each appearing in four layers. Separate symmetry groups were defined for each layer and also separate for Normal and Dark matter.

In the Fall of 2019 the author considered the possibility that the UST might follow from a deeper theory. He considered the possibility of a theory with a space defined by hypercomplex coordinates. In January, 2020 the author released a book describing such a theory based on complex octonion coordinates that he called QUeST. This theory had a fundamental fermion spectrum that was identical to the UST and a set of internal symmetries almost identical to the UST. This remarkable agreement motivated the author to continue to explore features of hypercomplex spaces. His first endeavor which is described in his January, 2020 book, Blaha (2020a), was to develop a hypercomplex space theory of the proposed Megaverse (Multiverse) called MOST.

Subsequently the present theory of hypercomplex spaces,[71] the HyperCosmos, was forged with working progress reports (due to high interest levels) presented in a series of books in 2020 and 2021. This book describes HyperCosmos theory in detail. It also describes a precursor theory the ProtoCosmos,[72] from which the HyperCosmos theory is derived.[73]

16.1 QUeST Origin in the HyperCosmos

In this section we describe the origin of QUeST and UST in the HyperCosmos spectrum and in the Fundamental Reference Frame.

16.1.1 The HyperCosmos Spectrum QUeST/UST Entry

The entry for the QUeST space of our universe is Blaha number 7 in the table of Fig. 12.1.

---

[71] The author was first led to develop a theory he called Octonion Cosmology, which recently was suyperceded by a simpler, yet deeper theory, he called the HyperCosmos.

[72] An initial form of Eos was presented in Blaha (2020a).

[7373] The author is not aware of any papers or books on UST, QUeST, MOST or related topics in these areas in the past years of effort.

Blaha Space Number	Cayley-Dickson Number	Cayley Number	Dimension Array column length	Space-time-Dimension	CASe Group $su(2^{r/2},2^{r/2})$
$N = o_s$	n	d_c	d_{cdN}	r	CASe
7	3	8	16	4	su(4,4)

The number of elements in the universe dimension array is 256 = d_{cd7}^2. This number gives the number of fundamental fermions in QUeST as well as the total number of symmetry group dimensions (including r = 4 space-time dimensions). QUeST is constructed from the N = 7 entry of the HyperCosmos table of spaces.

16.1.2 Fundamental Reference Frame Origin

At a deeper level we can see the contents and form of the fermion and symmetry spectrums from Fundamental Reference Frame considerations. Chapter 9 shows in detail the Fundamental Reference Frame features for Blaha number N = 7.

Fermions

There are 16 copies of the 16 fermions in a Fundamental Reference Frame as shown in Fig. 9.2 (reproduced below for the reader's convenience).

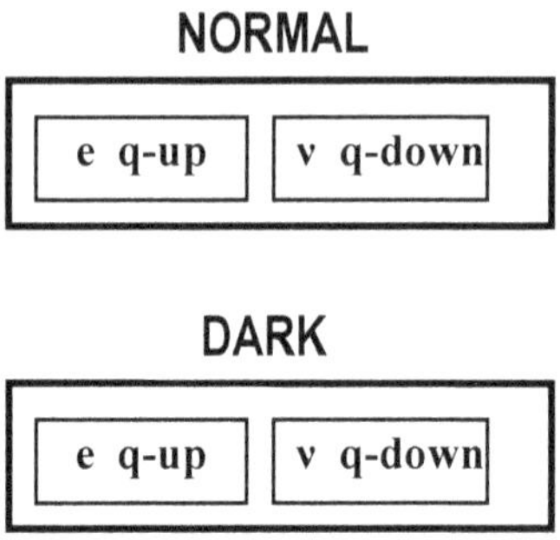

Figure 9.2. The 16 fermions that constitute the set that is duplicated 16 times in the 256 fermion spectrum for N = 7 space (universe).

The copies form 16 rows that we subdivide into four layers of four generations. As shown in Fig. 15.1. Each row contains 8 Normal fermions and 8 Dark fermions.

The separation of the 16 fermions into pairs of four fermions in Fig. 9.2 and the form of Fig. 9.1 reflects the separation of fermions and internal symmetries into 4, 8, 16 and 32 parts due to symmetry splitting that we discussed in Blaha (2021b).

Symmetry Groups

The set symmetry groups in QUeST for our universe also exhibits the same sort of duplication seen for fundamental fermions. The symmetry groups have a 256 real-valued dimensions representation as the product of the fundamental representations of their component groups.

The basic symmetry group set for N = 7 space is a U(4)⊗U(4) group occupying 16 representation dimensions in the Fundamental Reference Frame:

U(4)⊗U(4)

After 16-fold duplication the U(4) factors in the copies undergo transformations/breakdowns to

U(4) or SU(3)⊗U(1) or SU(2)⊗U(1)⊗SL(2, **C**)[74]

Initially the N = 7 full set of groups has the form of 16 U(4)⊗U(4) symmetry groups (Fig. 9.3.) Then they are transformed to the groups in Fig. 9.4. Fig. 9.3 is reproduced below for convenience.

	NORMAL	DARK
Layer 1	U(4)⊗U(4) U(4)⊗U(4)	U(4)⊗U(4) U(4)⊗U(4)
Layer 2	U(4)⊗U(4) U(4)⊗U(4)	U(4)⊗U(4) U(4)⊗U(4)
Layer 3	U(4)⊗U(4) U(4)⊗U(4)	U(4)⊗U(4) U(4)⊗U(4)
Layer 4	U(4)⊗U(4) U(4)⊗U(4)	U(4)⊗U(4) U(4)⊗U(4)

Figure 9.3. The "initial" distribution of sets of N = 7 symmetry groups. Each set is distinct and supports interactions only for the corresponding set of fermions (separately for Normal and Dark fermions) in Fig. 9.1. *Thus each set of 16 fermion generations has its own quantum numbers and interactions.* Each U(4)⊗U(4) set has a 16 real-valued dimension representation, which will be of importance when we consider Fundamental Reference Frames in section 9.3.

---

[74] We use SL(2, C) to represent $SO^{+}(1, 3)$.

After symmetry breaking one obtains:

NORMAL		DARK	
SU(3)⊗U(1) Generation U(4)	SU(2)⊗U(1)⊗SL(2, C) Layer U(4)	SU(3)⊗U(1) Generation U(4)	SU(2)⊗U(1)⊗SL(2, C) Layer U(4)
SU(3)⊗U(1) Generation U(4)	SU(2)⊗U(1)⊗SL(2, C) Layer U(4)	SU(3)⊗U(1) Generation U(4)	SU(2)⊗U(1)⊗SL(2, C) Layer U(4)
SU(3)⊗U(1) Generation U(4)	SU(2)⊗U(1)⊗SL(2, C) Layer U(4)	SU(3)⊗U(1) Generation U(4)	SU(2)⊗U(1)⊗SL(2, C) Layer U(4)
SU(3)⊗U(1) Generation U(4)	SU(2)⊗U(1)⊗SL(2, C) Layer U(4)	SU(3)⊗U(1) Generation U(4)	SU(2)⊗U(1)⊗SL(2, C) Layer U(4)

Figure 9.4. The transformed/broken sets of symmetries in QUeST (UST) and in N = 7 HyperCosmos space.. Note each element has a 16 real dimension representation. This depiction is also evident in QUeST and the UST.

There is a set of symmetries for each of the 4 layers separately for Normal and Dark symmetries. Each set occupies 32 dimensions. Thus 8•32 = 256 total symmetry dimensions.

16.2 Detailed Summary of UST, QUeST, NEWUST, and NEWQUeST

This section summarizes features of these four *equivalent* theories. The theories can be found in

Blaha, 2018e, *Unification of God Theory and Unified SuperStandard Model THIRD EDITION*

Blaha, 2020a, *Quaternion Unified SuperStandard Theory (The QUeST) and Megaverse Octonion SuperStandard Theory (MOST)*

Blaha, 2020c, *Unified SuperStandard Theories for Quaternion Universes & The Octonion Megaverse*

Blaha, 2021c, *Beyond Octonion Cosmology*

Blaha, 2021d, *Universes are Particles*

Blaha, 2021e, *Octonion-like dna-based life, Universe expansion is decay, Emerging New Physics*

Blaha, 2021f, *The Science of Creation New Quantum Field Theory of Spaces* as well as earlier books by the author.

The theories originated in the past twenty years from the Standard Model of Particles with SU(2)⊗U(1)⊗SU(3), and Two-Tier Quantum and PseudoQuantum Field Theory. Noting the presence of conserved particle numbers, and the presence of at least three fermion generations, we introduced the U(4) Generation Groups and the U(4) Layer Groups[75] together with four layers of four generation fermions. The "Normal" fermions had a corresponding set of four layers of four generations of "Dark" fermions. The result was the Unified SuperStandard Theory (UST) symmetry:

$$\{[SU(2)\otimes U(1)\otimes SU(3)]^2\otimes U(4)^4\}^4$$

with an additional Strong Interaction U(1) group (analogous to that of ElectroWeak theory) found to be needed. Space-time had four dimensions.

In January, 2020 the author discovered that an octonion-based theory that he constructed and named QUeST had the same internal symmetries as UST with the addition of $U(1)^8$ QUeST's internal symmetry, which could be based on a 16 × 16 dimension array, was

$$[SU(2)\otimes U(1)\otimes SU(3)\otimes U(1)]^8\otimes U(4)^{16}$$

The addition of $U(1)^8$ indicates that the Strong Interactions in the theory are broken Strong SU(4).

During 2020 the author developed an octonion space spectrum for both universes and Megaverses, and other spaces. The spectrum was shown to arise from a generation mechanism whereby fermion-antifermion annihilation in a higher space produced an instance of a lower space. A critical part of the derivation of the octonion spectrum was the realization that even space-time dimension spinor arrays are composed of Cayley number rows and columns. Spinor arrays of annihilating fermion-antifermion pairs were shown to generate the dimension arrays of subspace instances. Analyzing the spinor arrays the author noted that the dimension array could be viewed as composed of 64 dimension blocks, which were further subdivided into 16 dimension subblocks.

The subblock structuring, using the known contents of the Standard Model plus Generation and Layer groups, gives the dimension array structure containing 4 × 4 subblocks in Figs. 16.1 and 16.2.

Thus there was a *most* satisfactory match between UST and QUeST with the only significant difference being the space-time: four octonion (complex quaternion) coordinates for QUeST and four real space-time coordinates for UST.

The form of the square spinor arrays generated by fermion-antifermion annihilation gives 64 dimension blocks and 16 dimension blocks as well as 32 dimension composite blocks that are evidenced in the NEWQUeST (and NEWUTMOST) fermion spectrums and internal symmetry group structure.

---

[75] See Appendix 16-A.

Since we see only real dimensions in Reality, we recently transferred 28 QUeST dimensions from space-time to $U(2)^7$ internal symmetry dimensions. The set of internal symmetries was increased by $U(2)^7$, which we call Connection Groups.[76] Each Connection group specifies interactions between corresponding fermions (e with e, q with q, and so on) in separate layers and between Normal and Dark fermions. The connections between the various blocks of fermions are shown in Fig. 16.3. *We implement the very practical rule that all blocks must be connected by interactions or they would not be of physical interest. A totally isolated block effectively does not exist physically (except possibly for gravitation effects).*

The interactions of the Connection groups must be very weak and/or their gauge bosons must be very massive.

The addition of the Connection Groups and the reduction of space-time dimensions accordingly results in NEWQUeST and NEWUST as summarized below. See appendix 16-A for a discussion of NEWQUeST Generation and Layer groups.

Note: the Generation, Layer, and Connection groups are all badly broken. Their vector bosons must be very massive since they have not been detected in experiments.

16.2.1 Internal Symmetries

The groups are ElectroWeak SU(2)⊗U(1), Strong SU(3), Generation Group U(4), Layer Group U(4), and U(2) and U(4) Connection groups obtained by transfer from space-time coordinates (See Blaha 2012c). The SU(3)⊗U(1) symmetry may be a broken SU(4) symmetry in eqs. 16.1 – 16.4. The internal symmetries for the theories are:

<u>**UST**</u>

$$[SU(2)\otimes U(1)\otimes SU(3)]^8\otimes U(4)^{16} \tag{16.1}$$

<u>**QUeST**</u>

$$[SU(2)\otimes U(1)\otimes SU(3)\otimes U(1)]^8\otimes U(4)^{16} \tag{16.2}$$

<u>**NEWQUeST**</u>

$$[SU(2)\otimes U(1)\otimes SU(3)\otimes U(1)]^8\otimes U(4)^{16}\otimes U(2)^7 \tag{16.3}$$

The only change is in internal symmetries: Twenty-eight real dimensions transferred from space-time coordinates to Connection group $U(2)^7$ internal symmetry.

<u>**NEWUST**</u>

$$[SU(2)\otimes U(1)\otimes SU(3)\otimes U(1)]^8\otimes U(4)^{16}\otimes U(2)^7 \tag{16.4}$$

The only change in internal symmetries: Twenty-eight real dimensions added for $U(2)^7$ Connection group internal symmetry.

---

[76] See section 11-A.4.

16.2.2 Space-Time Coordinates

UST

Four real space-time coordinates.

QUeST

Four octonion (complex quaternion) coordinates.

NEWUST

Four real space-time coordinates. No change from UST space-time.

NEWQUeST

Four real space-time coordinates. The six coordinates in the n = 4 octonion space (Figs. 1.5 and 1.6) were lowered to four space-time coordinates with two coordinates transferred to Connection groups.

The only change is in space-time coordinates: Fourteen dimensions transferred from QUeST space-time coordinates to Connection group $U(2)^7$ internal symmetry.

16.3 Fundamental Fermion Spectrum

There are 256 fundamental fermions[77] in NEWQUeST and NEWUST. Conceptually their structure can be viewed as an extrapolation of the known three generations of The Standard Model. For reasons given previously and in Appendix 16-A a fourth generation was indicated and a corresponding Dark sector of similar structure was added. In addition, because of the need for Layer groups (Appendix 16-A), the overall structure consisted of four copies of this layer.

Correspondingly, each layer also has its own set of internal symmetry gauge groups[78] to limit mixing between the layers to Layer group interactions and Connection group interactions.

Fig. 16.4 shows the structure of the NEWQUeST/NEWUST fermions. The blocks are 4 × 4 reflecting the origin of the NEWQUeST/NEWUST space (universe) instance from Megaverse fermion-antifermion annihilation. The spinor analysis of their spinor arrays yields a 16 dimension-based block structure. The 64 dimension fermion layers reflect the 64 dimension structuring of the Megaverse obtained from its creation by fermion-antifermion creation in the Maxiverse. . .

16.4 Total Dimensions

The total of internal symmetry and space-time dimensions is 256 in all four theories listed above. It is based on the 16 × 16 dimension array of the Cayley number n = 4 space of the Octonion spectrum.

16.5 Pattern of Internal Symmetries

The NEWQUeST dimension array for internal symmetries is subdivided into four layers of 56 dimensions—just as in NEWUST (and UST). Fig. 16.1 displays the layers using SU(4) in place of SU(3)⊗U(1). Each layer has a block of 28 dimensions

---

[77] See Fig. 15.1 for the fermion spectrum determined by a spinor array.
[78] See Fig. 15.3 for the pattern of symmetry groups.

for Normal and 28 dimensions for Dark sectors. There are also seven U(2) Connection groups[79] plus four real-valued space-time coordinates bringing the NEWQUeST total to 256 dimensions.= 4*56 + 28 + 4 = 256 dimensions. The Connection groups are shown in Fig. 16.2 (and section 16-A.4), which is a revision of the pattern shown in Blaha (2021c).

[79] See Appendix 11-A.

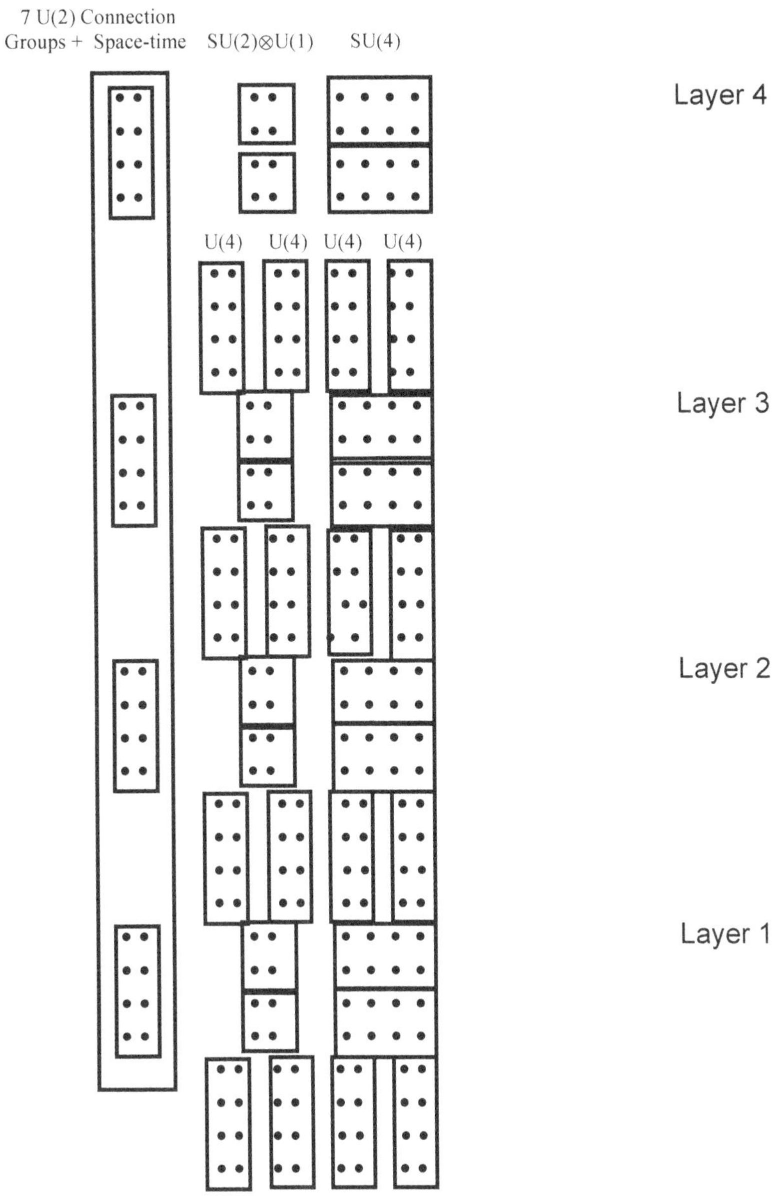

Figure 16.1. The four layers of NEWUST and NEWQUeST internal symmetry groups (and space-time) with SU(4) before breakdown to SU(3)⊗U(1). Note the left column of blocks combine to specify a 4 dimension real space-time plus seven U(2) Connection groups. Note each layer has 64 dimensions = 56 + 8 dimensions.

Layers	NORMAL		DARK	
	4	4	4	4
4	SU(2)⊗U(1)⊗SU(3)⊗U(1) 4 Space-time Dimensions	Generation + Layer Groups	SU(2)⊗U(1)⊗SU(3)⊗U(1) 4 Space-time Dimensions	Generation + Layer Groups
4	SU(2)⊗U(1)⊗SU(3)⊗U(1) 4 Space-time Dimensions	Generation + Layer Groups	SU(2)⊗U(1)⊗SU(3)⊗U(1) 4 Space-time Dimensions	Generation + Layer Groups
4	SU(2)⊗U(1)⊗SU(3)⊗U(1) 4 Space-time Dimensions	Generation + Layer Groups	SU(2)⊗U(1)⊗SU(3)⊗U(1) 4 Space-time Dimensions	Generation + Layer Groups
4	SU(2)⊗U(1)⊗SU(3)⊗U(1) 4 Space-time Dimensions	Generation + Layer Groups	SU(2)⊗U(1)⊗SU(3)⊗U(1) 4 Space-time Dimensions	Generation + Layer Groups

Figure 16.2.. Four layers of Internal Symmetry groups in NEWQUeST (omitting Connection groups). The groups in each layer are independent of those in other layers. The groups in each block of each layer are independent of those in the other blocks. Each block contains 16 dimensions. The block dimensions furnish fundamental representations for the groups listed. The entire set of blocks contains 256 dimensions. Each layer contains 56 internal symmetry dimensions. The first two columns are for the "Normal" sector. The last two columns are for the "Dark" sector (although most of the Normal sector is Dark observationally at present.) This figure also holds for UST with the addition of U(1) groups. The eight sets of 4 real dimension space-times combine to give a 4 real dimension space-time and seven U(2) Connection groups.

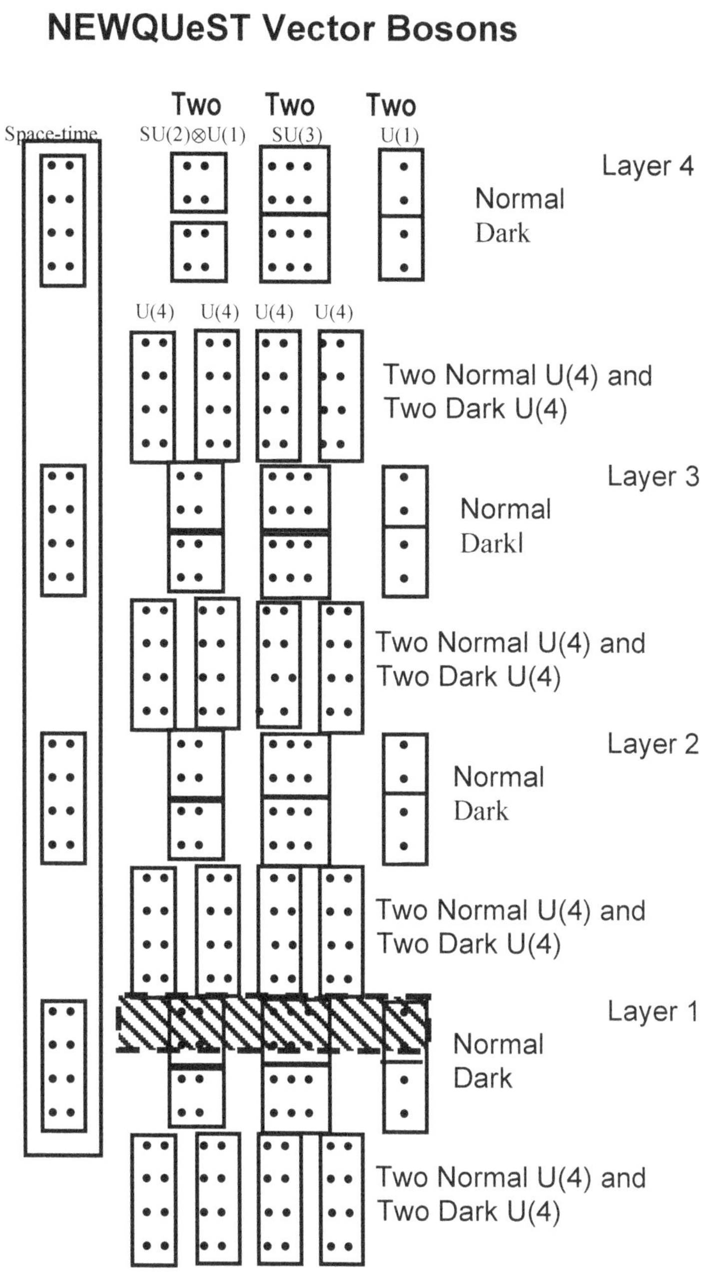

Figure 16.2a. The four layers of QUeST internal symmetry groups (and space-time) for the 32 octonion dimension form of space. Note: each row has an 8 • octonion. Note the left column of blocks combine to specify a 4 dimension octonion space-time and seven U(2) Connection Groups. Note each layer requires 64 dimensions.. *Note the duplication (using the "Two" label) of each symmetry in each layer. The cross hatched area indicates the known symmetry groups.*

Connection Group Applied to Fermions

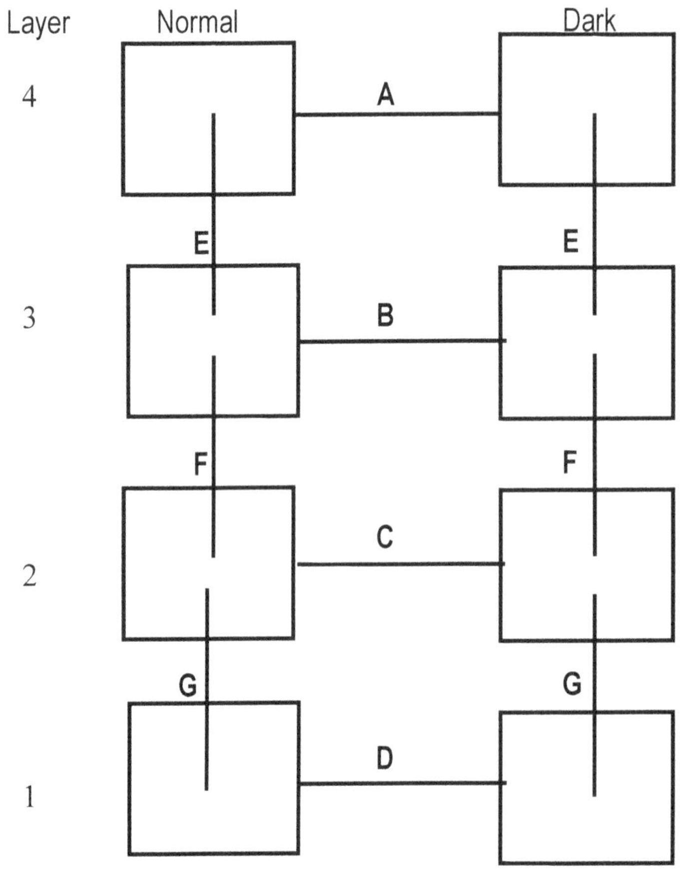

Figure 16.3. The seven U(2) Connection groups (shown as 10 lines) between the eight NEWQUeST/NEWUST blocks. Connection groups are obtained by transfering 28 dimensions from QUeST space-time to internal symmetries with the consequent reduction of the space-time from four octonion (complex quaternion) coordinates to four real coordinates. The Connection groups generate rotations and interactions between corresponding fermions and vector bosons of each pair of blocks. See Appendix 11-A. The Normal and Dark sector U(2) vertical connections above (E, F, G) represent the same U(2) groups.

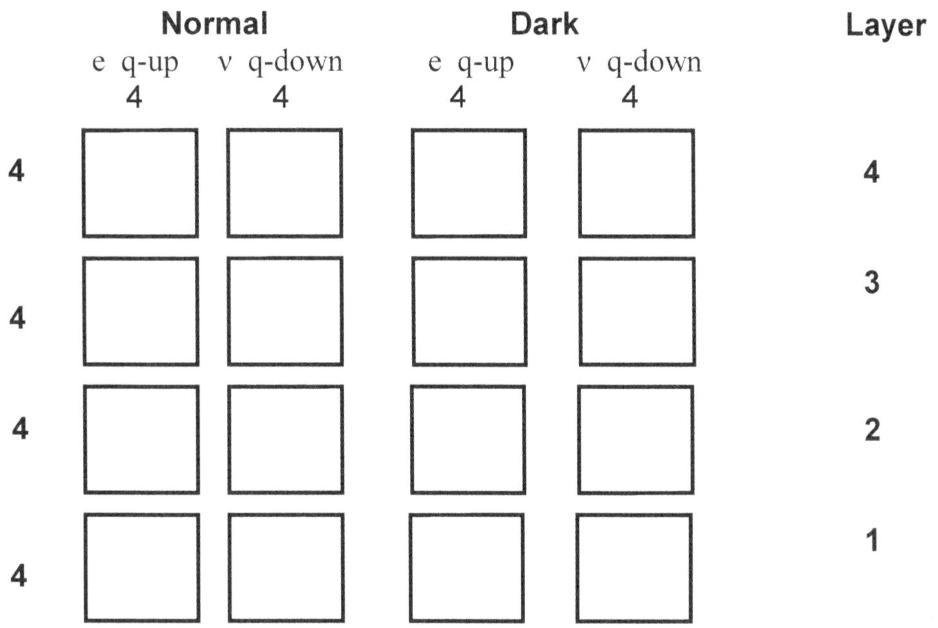

Figure 16.4. Block form of a 16 × 16 NEWQUeST/NEWUST fermion array with each block row corresponding to one layer. Each block contains four generations of fermions. The result is 4 × 4 blocks. The label e q-up indicates a charged lepton – up-type quark pair, ν q-down indicates a neutral lepton – down-type quark pair, and so on. The blocks can be viewed as SU(3)⊗U(1) or broken SU(4) blocks.

Appendix 16-A Generation and Layer Groups

16-A.1 U(4) Generation Groups

In the Big Bang all particles were massless and all symmetries unbroken. Hence the four Normal particle number symmetries, and the four Dark particle number symmetries, are all "conserved" in the Big Bang. Afterwards conservation laws are then broken.

We define two particle number operators for normal up-quark particles and down-quark particles, B_{uq} and B_{dq}. Similarly we define two particle number operators for normal species "e" (electron) particles and species "ν" particles, B_e and B_ν. Similarly we define Dark matter equivalents:[80] B_{De}, $B_{D\nu}$, B_{Duq}, and B_{Ddq}.

In the absence of symmetry breaking these fermion particle number operators would be conserved. Thus there are two sets of "diagonal" operators with associated U(4) groups for the Normal and Dark sectors. They are part of the Normal U(4) Generation Group and the Dark U(4) Generation Group.

The fermion fundamental representation of a U(4) group has four fermions. U(4) has rotations, and also interactions of the form $\overline{\Psi}\gamma\cdot B\cdot T\Psi$ where Ψ is a fermion four-vector, B is a 16 component U(4) gauge field, and T consists of 16 component 4×4 U(4) arrays.

In the case of the Generation Group the gauge fields have electric charge zero. Since the four species have different electric charges (1, 0, 2/3, -1/3) the U(4) gauge boson fields cannot mix the fermions of different species. Generation Group interactions are diagonal[81] in fermion species (e, ν, up-quark, and down-quark species).

Consequently the U(4) Generation Group must have a reducible representation D consisting of a set of four fundamental U(4) representations, D_e, D_ν, D_{upq}, and D_{dnq}, appearing in blocks along the diagonal of D. Each block is a separate U(4) irreducible representation for a species (due to the electric charge superselection rule.)

There are four generations of each species in the Normal and in the Dark matter sectors. The four generations for each fermion species: e, ν, up-quark, and down-quark each furnish a U(4) fundamental representation within the reducible representation D. The fourth generation of normal fermions has not as yet been found due to their extremely large masses.

The Generation Group rotates the fundamental fermions of each fundamental representation separately for each of the four species of each of the four layers.[82] Thus

---

[80] By analogy, we assume that there are four species of Dark matter: charged Dark leptons, neutral Dark leptons, Dark up-type quarks, and Dark down-type quarks. Thus we are led to the Dark particle numbers: Dark Baryon Numbers, and Dark Lepton Numbers shown above.

[81] ElectroWeak interactions can cross between species due to their charged gauge vector bosons.

[82] There are separate Generation groups for each layer.

the Generation Group guarantees that all generations of each species have the same electric charge and other quantum numbers.

The U(4) Generation Group also specifies a gauge field interaction among the fermions of its fundamental representation, species by species, for both Normal and Dark sectors. The form of the interactions for the Normal sector for each fermion layer is:

$$g_e\Psi_e\gamma\cdot B_e\cdot T\Psi_e + g_v\overline{\Psi}_v\gamma\cdot B_v\cdot T\Psi_v + g_{upq}\overline{\Psi}_{upq}\gamma\cdot B_{upq}\cdot T\Psi_{upq} + g_{dnq}\overline{\Psi}_{dnq}\gamma\cdot B_{dnq}\cdot T\Psi_{dnq} \quad (16\text{-A.1})$$

where g_e … are coupling constants, the gauge vector fields are B_e … , and the Ψ_e … are 4-vectors of fermions of the four generations of each species in a layer.

The gauge vector bosons of the Generation Group have large masses. If the conservation of the fermion particle numbers is broken then we view it as a consequence of Generation Group symmetry breaking.

Generation Group rotations guarantee the internal quantum numbers of each generation of each species are the same since symmetry breakdown is not present at the instant of the Big Bang.

The above discussion applies similarly to the Dark sector. There are 8 Generation Groups in total in NEWQUeST/NEWUST.

16-A.2 U(4) Layer Groups

The set[83] of particle number operators can be extended if we take account of the fourfold fermion generations.

We can subdivide the above particle number sets into four additional particle numbers *per generation*. For the ith generation (of the four generations) we define

L_{ie} – The "e" species particle number for the ith generation
L_{iv} – The ν species particle number for the ith generation
L_{iuq} – The up-quark species particle number for the ith generation
L_{idq} – The down-quark species particle number for the ith generation

L_{iDe} – The Dark "e" species particle number for the ith generation
L_{iDv} – The Dark ν species particle number for the ith generation
L_{iDuq} – The Dark up-quark species particle number for the ith generation
L_{iDdq} – Dark down-quark species particle number for the ith generation

for each generation i = 1, 2, 3, 4. Individual fermions have positive $L_{ia} = +1$ values and antifermions have negative $L_{ia} = -1$ values for each species.

At this point we have a set of four particle number operators for each of four generations (i = 1, 2, 3, 4) of fermions in the Normal sector and similarly in the Dark sector. We then define a U(4) group framework for each set of particle numbers.

---

[83] Here again, in the Big Bang all particles were massless and all symmetries unbroken. Hence particle numbers are "conserved" in the Big Bang. Conservation is then broken afterwards in most cases.

The only way to specify fundamental representations for each of the four sets in a sector is to assume there are four layers, with each layer having four generations, and with a fundamental U(4) representation defined for each generation composed of fermions from each layer. Thus there are four Layer Groups for each Normal and each Dark sector: a Layer Group for generation 1, a Layer Group for generation 2, and so on.

The Layer Groups are also "split" by species due to the electric charge superselection rule. Each Layer Group is diagonal in the four fermion species. All their gauge fields are electrically neutral.

Consequently each of the four U(4) Layer Groups in the Normal fermion sector has a reducible U(4) representation D_j for j = 1, 2, 3, 4. Each reducible representation is composed of four irreducible U(4) representations for each species due to the electric charge superselection rule:

$$D_j = D_{je} + D_{j\nu} + D_{jupq} + D_{jdnq},$$

for j = 1, 2, 3, 4.

There are four layers of each species in the Normal and in the Dark matter sectors. The second, third and fourth layers of normal fermions has not as yet been found due to their extremely large masses.

A Layer Group rotates the fundamental fermions of each fundamental representation separately for each of the four species of each of the four generations.

The Layer Groups guarantee that all layers of each species have the same electric charge and other quantum numbers.

Each U(4) Layer Group also specifies a gauge field interaction among the fermions of its fundamental representation, species by species, for both Normal and Dark sectors. The form of the interactions is:

$$g_{ei}\overline{\Psi}_{ei}\gamma\cdot C_{ei}\cdot T\Psi_{ei} + g_{\nu i}\overline{\Psi}_{\nu i}\gamma\cdot C_{\nu i}\cdot T\Psi_{\nu i} + g_{upqi}\overline{\Psi}_{upqi}\gamma\cdot C_{upqi}\cdot T\Psi_{upqi} + g_{dnqi}\overline{\Psi}_{dnqi}\gamma\cdot C_{dnqi}\cdot T\Psi_{dnqi} \qquad (16\text{-A.2})$$

for i = generation = 1, … , 4, where g_{ei} … are coupling constants, the gauge fields are C_{ei} … , and the Ψ_{ei} … are 4-vectors of fermions formed of the ith generation fermions in each layer of each species.

The gauge vector bosons of the Layer Groups also have large masses. If the conservation of the fermion particle numbers is broken then we view it as a consequence of Layer Groups symmetry breaking.

Layer Group rotations guarantee the internal quantum numbers of each layer of each species are the same since symmetry breakdown is not present at the instant of the Big Bang.

The above discussion applies similarly to the Dark sector. There are 8 of Layer Groups in NEWQUeST/NEWUST.

Fig. 15.1 shows the fundamental fermion spectrum with the representations of the Generation groups and Layer groups indicated.

Experimentally, we know of three generations of fermions—the lowest 3 generations of the lowest level. The remaining 4th generation and three layers of fermions are of much higher mass and are yet to be found.

See Blaha (2019g) and (2018e) for a detailed discussion of the Layer Groups. We note in passing that the symmetries of these number operators are badly broken. Yet the underlying group structure remains.

17. NEWUTMOST For the Megaverse

In this section we describe the N = 6 HyperCosmos space for the Megaverse (Multiverse). This space was called MOST in 2020 and thence after some consideration, and minor changes, became UTMOST and NEWUTMOST recently.

17.1 The HyperCosmos Spectrum NEWUTMOST Entry

The entry for the NEWUTMOST space of the Megaverse is Blaha number 6 in the table of Fig. 12.1.

Blaha Space Number $N = o_s$	Cayley-Dickson Number n	Cayley Number d_c	Dimension Array column length d_{cd}	Space-time-Dimension r	CASe Group $su(2^{r/2},2^{r/2})$ CASe
6	4	16	32	6	su(8,8)

The number of entries in the universe dimension array is $d_{cd7}^{\ 2} = 32^2 = 1024$. This number gives the number of fundamental fermions in NEWUTMOST as well as the total number of symmetry group dimensions (including r = 6 space-time dimensions).

17.1.1 Fundamental Reference Frame Origin

NEWUTMOST is constructed from the N = 6 entry of the HyperCosmos table of spaces. The fermions are quadrupled to 1024 fermions with the form of the NEWUTMOST fermion spectrum consisting of four copies of the N = 7 fermion spectrum depicted in Fig. 15.1. The 32 copies become 8 layers of four fermion generations. Fig. 17.1 shows the form of the NEWUTMOST fermion spectrum.

17.1.2 Fundamental Reference Frame Origin

The contents and form of the fermion and symmetry spectrums can be derived from Fundamental Reference Frame considerations. Chapter 9 shows indicates the Fundamental Reference Frame features for Blaha number N = 6.

Fermions

There are 32 copies of the 32 fermions in a Fundamental Reference Frame as shown in Fig. 9.5 and 9.6.

4	4	4	4	4	4	4	4
e q-up	v q-down	e q-up	v q-down	e q-up	v q-down	e q-up	v q-down

These 32 fermions are transformed to 32 copies in a static frame. The resulting 1024 fermion spectrum has the form of Fig. 17.1 with the NEWUTMOST dimension array consisting of four copies of the N = 7 dimension array. Fig. 17.2 shows the fermion array in more detail with 32 row copies of the 32 fermions in the N = 6 Fundamental Reference Frame.

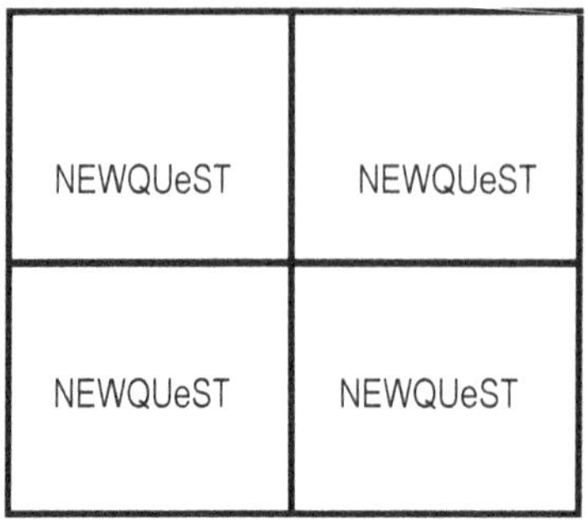

Figure 17.1. The NEWUTMOST dimension array viewed as composed of four copies of NEWQUeST.

<u>Symmetry Groups</u>

The set symmetry groups in NEWUTMOST exhibits the same sort of duplication seen for fundamental fermions. The symmetry groups have a 1024 real-valued dimensions representation as the product of the fundamental representations of their component groups. They are

$$\{[SU(2)\otimes U(1)\otimes SU(3)\otimes U(1)]^8\otimes U(4)^{16}\otimes U(2)^7\}^4\otimes SU(4)\otimes U(1)\otimes SO(1, 5) \tag{12.1}$$

where $U(2)^7\}^4\otimes SU(4)\otimes U(1)$ are Connection groups discussed in Chapter 11, section 11.3. The group SO(1, 5) is the Lorentz group generalization to 6 dimension space-time.

The basic symmetry group set for N = 6 space is are two $U(4)\otimes U(4)$ groups occupying 32 real-valued dimensions in the Fundamental Reference Frame:

$$U(4)\otimes U(4) \quad U(4)\otimes U(4)$$

They are transformed 32 copies in a static reference frame. The 32 copies may be viewed as four copies of the N = 7 set of symmetries shown in Fig. 9.3 below.

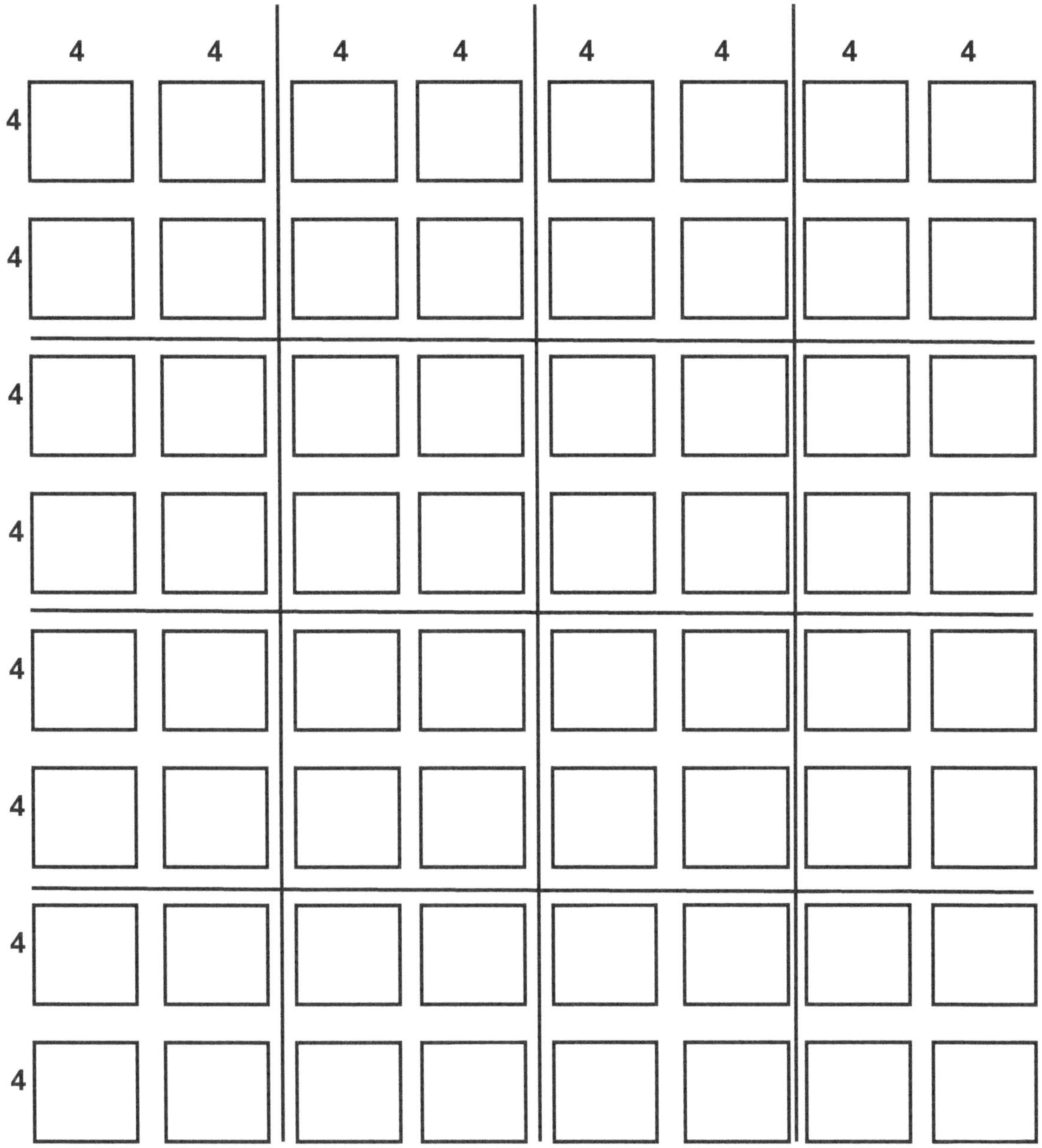

Figure 17.2. Four layers (each occupying two rows) in the 32 × 32 dimension NEWUTMOST array.

Each U(4)⊗U(4) blocks is initially transformed to

SU(2)⊗U(1)⊗SL(2,C)⊗SU(4)

or remains

U(4)⊗U(4)

blocks by symmetry breakdown in the output set of symmetries. Four block sets of symmetries appear in four layers. Fig. 9.4 below shows one of the four copies of the QUeST N = 7 set residing in the NEWUTMOST set of symmetries.

	NORMAL	DARK
Layer 1	U(4)⊗U(4) U(4)⊗U(4)	U(4)⊗U(4) U(4)⊗U(4)
Layer 2	U(4)⊗U(4) U(4)⊗U(4)	U(4)⊗U(4) U(4)⊗U(4)
Layer 3	U(4)⊗U(4) U(4)⊗U(4)	U(4)⊗U(4) U(4)⊗U(4)
Layer 4	U(4)⊗U(4) U(4)⊗U(4)	U(4)⊗U(4) U(4)⊗U(4)

Figure 9.3. The "initial" distribution of sets of N = 7 symmetry groups. Four copies appear in NEWUTMOST. Each set is distinct and supports interactions only for the corresponding set of fermions (separately for Normal and Dark fermions) in Fig. 9.1. *Thus each set of 16 fermion generations has its own quantum numbers and interactions.* Each U(4)⊗U(4) set has a 16 real-valued dimension representation, which will be of importance when we consider Fundamental Reference Frames in section 9.3.

After symmetry breaking one initially (before the Connection groups) obtains:

NORMAL		DARK	
SU(3)⊗U(1) Generation U(4)	SU(2)⊗U(1)⊗SL(2, C) Layer U(4)	SU(3)⊗U(1) Generation U(4)	SU(2)⊗U(1)⊗SL(2, C) Layer U(4)
SU(3)⊗U(1) Generation U(4)	SU(2)⊗U(1)⊗SL(2, C) Layer U(4)	SU(3)⊗U(1) Generation U(4)	SU(2)⊗U(1)⊗SL(2, C) Layer U(4)
SU(3)⊗U(1) Generation U(4)	SU(2)⊗U(1)⊗SL(2, C) Layer U(4)	SU(3)⊗U(1) Generation U(4)	SU(2)⊗U(1)⊗SL(2, C) Layer U(4)
SU(3)⊗U(1) Generation U(4)	SU(2)⊗U(1)⊗SL(2, C) Layer U(4)	SU(3)⊗U(1) Generation U(4)	SU(2)⊗U(1)⊗SL(2, C) Layer U(4)

Figure 9.4. The transformed/broken sets of symmetries in QUeST (UST) and in N = 7 HyperCosmos space.. Four copies appear in NEWUTMOST. Note each element has a 16 real dimension representation. This depiction is also evident in QUeST and the UST.

The split into four N = 7 copies includes a set of 7 Connection groups in each copy and SU(4)⊗U(1) Connection groups joining the four copies together. See Figs. 11.3 and 11.4. These figures implement the rule that Connection groups maximally join the four parts of the fundamental fermion spectrum.

18. Other Universes

The other spaces (universes) of the HyperCosmos each have a structure that can be viewed as building through quadrupling from the next lower space. This is ultimately a result of the Cayley numbers at the base of HyperCosmos spectrum structure.

The Blaha number $N \leq 6$ spaces are:

Blaha Space Number	Cayley-Dickson Number	Cayley Number	Dimension Array column length	Space-time-Dimension	CASe Group $su(2^{r/2},2^{r/2})$
$N = o_s$	n	d_c	d_{cd}	r	CASe
0	10	1024	2048	18	su(512,512)
1	9	512	1024	16	su(256,256)
2	8	256	512	14	su(128,128)
3	7	128	256	12	su(64,64)
4	6	64	128	10	su(32,32)
5	5	32	64	8	su(16,16)
6	4	16	32	6	su(8,8)

18.1 The Quadrupling Process

In each space the dimension array is the quadruple of the dimension array of the next lower space. The quadrupling process is illustrated by Figs. 17.1, 15.4, and 11.4.

The benefit of quadrupling appears in the simplification of the dimension array, the fermion spectrum, and the set of symmetries.

18.2 Fermion Quadrupling

The quadrupling process causes the fermion spectrum of a number N space to be quadruple that of the next higher space with Blaha number N + 1. The number of fermions in a space is $d_{dN} = 2^{22-2N}$.

18.3 Symmetry Group Quadrupling

The quadrupling process causes the number of dimensions allocated to symmetry representations in the dimension array of a number N space to quadruple in the next higher space with Blaha number N + 1. The number of symmetry dimensions in a space is $d_{dN} = 2^{22-2N}$. The groups in Blaha number N space for $N \leq 7$ are

$$\{[SU(2)\otimes U(1)\otimes SU(3)\otimes U(1)\otimes U(4)^2]^{64-8N}\otimes G_{ConnectionsN}\otimes SO(1, r - 1) \tag{18.1}$$

where $G_{ConnectionsN}$ is the product of the Connection Groups.

19. Unification of General Relativity and Quantum Theory

The unification of General Relativity and Quantum Theory is a widely discussed issue. The meaning of the word "unification" is not entirely clear. There are several possible meanings:

1. A "deep" unification where one symmetry and/or one Lagrangian (or equivalent) determines all particles and interactions. SuperString theories have this goal.

2. A unification based on the assembly of all symmetry groups into one comprehensive symmetry, and all interactions into one dynamic expression that implies all dynamical equations. HyperCosmos theory has implemented this possibility to create a unified theory as shown in this chapter.

This chapter describes HyperCosmos unification in some detail.

19.1 HyperCosmos Unification Overview

The HyperCosmos may be viewed as one large symmetry group SU(2048) that is broken to the set of spaces, including the ten physical spaces that support the creation of universes and appear in Fig. 1.2. The symmetry groups of the ten spaces each support internal symmetries and a Lorentz group for space-time coordinates. Thus internal symmetries and space-time symmetry are united in each space.

A term in a Riemann-Christoffel curvature appears for each internal symmetry group and General Relativity. The particles of the space have a quantum Lagrangian formulation that supports the creation of dynamical equations and that support a divergence-free perturbation theory in any of the ten spaces. Gravitation has a quantum field theory formulation in the weak field expansion that serves the purpose of having a Quantum Gravity since space is almost flat in any small neighborhood of a space-time except singular points. In the small, Quantum Gravity is well-defined with no divergences using the author's Two-Tier Quantum Field Theory (Appendix A).

Thus we have unification in the HyperCosmos.

19.2 Unification Details

The unification of General Relativity and Quantum Theory has been a goal of Physics for approximately one hundred years. The meaning, and process, of unification has been the subject of discussion for almost as long. This book formulates a unification procedure and then proceeds to examine the ingredients and form of unification. We will see that The HyperCosmos contains the essential ingredients for unity: a combination of space-times with internal symmetries in each of its spaces (universes.)

The form of General Relativity, in relation to the form of other particle interactions, reveals a unity based on GiFT. A careful analysis of elementary particle field theory in differing coordinate systems leads to the Generalized Field Theory's (GiFT) Creation/Annihilation Space (CASe), which furnishes a further basis for unification with General Relativity.

19.3 The Form of General Relativity

General Relativity was first formulated by Einstein based on the Classical Physics (differential equations) of coordinate systems. It appeared different from the gauge field formulation of internal symmetries. As a result there seemed to be a gap between General Relativity and Elementary Particle Theory. Attempts were made to bridge the gap by quantizing General Relativity directly. These attempts seem to have not changed the view that General Relativity and Elementary particle theory were inherently different.

A different approach within a *vierbein* framework appeared due to the work of T. Kibble and others. This approach led to a gauge field formulation of General Relativity Thus this type of formulation of General Relativity places it on the same basis as the formulation of internal symmetry gauge fields.

A significant issue for unification is the appearance of infinities in gravitation perturbation theory. These infinities do not appear when Gravitation perturbation theory is done within the framework of the Two-Tier Theory within GiFT.

19.4 The HyperCosmos Framework

The similarity between the *vierbein* formulation of General Relativity and elementary particle internal symmetry gauge theory raises the question of origin: Is there a common origin?

The HyperCosmos provides a common origin in each of its ten universes (spaces). Part of each space has coordinates for each fundamental irreducible group representation. Part of each space is for space-time coordinates. The transformation properties of the space-times lead to a General Relativity for each universe. The transformations between static frames and a Fundamental Reference Frame further support the reality of unification.

Both the General Relativity and the internal symmetries can be formulated as gauge theories. Are there relations between coupling constants? Yes, but they differ in principle from universe to universe. Is there an ultimate unification value for all coupling constants in a universe? There may be such a unification value. But one must also consider the partition of internal symmetries in each universe due to the form of spaces as we described in previous work. Partitioning may obviate the value of a common coupling constant value.

19.5 Quantization Process

The process of quantization begins with the issue of particle quantization vs. wave quantization: Should matter and energy be quantized as particles or as waves? The answer in particle quantization is simple if one wishes creation and annihilation events.

Particles can interact to produce other particles in a "billiard ball" picture. Waves require complex differential equations at best.

Given a particle formulation it makes sense to define particle creation, a, and annihilation operators, $a^\dagger$, that can embody interactions that can change a set of incoming particles to a set of outgoing particles. At this point, if we introduce a "momentum" α for each particle, the question arises: Can particles have the same momentum? The Pauli Exclusion Principle states that fermion particles cannot. So we can set $(a^\dagger)^2 = 0$ for fermions. Bosons are not constrained by the Pauli Exclusion Principle so $(a^\dagger)^2 \neq 0$. One concludes the particle definition concepts by defining field operators for particles with Fourier expansions in α. From particle quantization one progresses to quantum fields, which in turn lead to Quantum Mechanics.

Having defined fields for particles in a coordinate system, then we note that a particle in that coordinate system becomes a superposition of particles when viewed in another coordinate in general. Einstein encountered this problem of exploring dynamic equations in one coordinate system in comparison with their equivalent in other coordinate systems. He resolved the problem with General Relativity.

19.6 Quantization in Differing Coordinate Systems

In our earlier work we showed that transformations between coordinate systems, where one or both coordinate systems do not have a Killing vector, are questionable in ordinary quantum field theory. We then showed quantization via GiFT provides a clear definition of particle states and their relation in the respective coordinate systems. (We again pursue the GiFT approach further for bosons and fermions in this book.)

The GiFT formulation of field theory is based on transformations between coordinate systems as is General Relativity. Thus both General Relativity and GiFT support the unity of General Relativity and Quantum Theory. General Relativity performs for coordinate systems what GiFT does for quantization in coordinate systems. Fig. 19.1 shows the correspondence between General Relativistic coordinate transformations and GiFT transformations in the space of creation/annihilation operators. A General Relativistic transformation between coordinate systems with Killing vectors may not induce a GiFT transformation. A General Relativistic transformation between coordinate systems where one coordinate system has a Killing vector, and the other does not, does induce a non-trivial GiFT transformation.

As we pointed out in earlier work GiFT embodies Quantum Theory in its entirety: PseudoQuantum Theory, Two-Tier Theory, Quantum Field Theory, and Quantum Mechanics. GiFT supports CQ Mechanics which contains Quantum Mechanics and Classical Mechanics. Thus we see the unity in General Relativity and Quantum Theory.

19.7 GiFT Creation/Annihilation Spaces

GiFT has creation/annihilation spaces created with CASe transformations. In chapter 4 we showed that creation/annihilation operators can form a basis in Blaha number N = 7 space with a 16-dimension space, a CASe space, which has a metric that is invariant under su(4,4) transformations of fermion particles.

The structure of CASe spaces and their group transformations parallels that of Special Relativity and General Relativity in space-time.

Thus General Relativity and GiFT Quantum Theory, which contains Quantum Field Theory and Quantum Mechanics, closely parallel each other in the features of transformations between coordinate systems, and in the gauge field format of interactions.

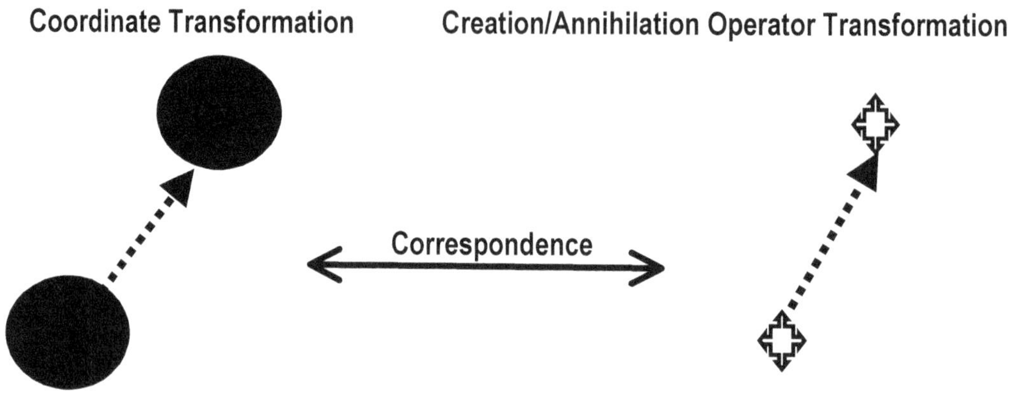

Figure 19.1. General Coordinate Transformations parallel Creation/Annihilation Operator Transformations.

19.8 CASe and Unification

The above considerations show the unity of General Relativity and Quantum Theory in our formulation. We find a unifying basis for General Relativity and Quantum Theory in considerations of transformations between coordinate systems, and in the existence of the relation between space-time transformations and CASe transformations. *Vierbein* gravity and the local gauge theory of internal symmetries exhibit a similar formulation. The structure of an overall fully unified theory of General Relativity and GiFT is evident.

In this book we developed a new view based on the CASe spaces of the GiFT formulation of Quantum theory and General Relativity. In this view HyperCosmos spaces are generated from CASe group transformations. CASe group transformations embody implicit General Relativistic transformations among static and non-static coordinates systems (reference frames). CASe transformations transform Creation and Annihilation operators, which implement Quantum theory at the most basic level. Thus the unification within GiFT.

At the deepest level the fundamental requirements for a theory of Physics are:

1. A particle formulation of matter and energy that leads to Quantum Theory.
2. A set of coordinate systems that embody General Relativity to ensure that physical processes are equivalent in the various coordinate systems: General Covariance.

As we have discussed previously, particle creation/annihilation operators are embodied in quantum field expansions; quantum fields require Quantum Field Theory; and Quantum Field Theory is the basis of *physical* Quantum Mechanics.[84]

The formulation of Quantum Field Theory within the GiFT PseudoQuantum framework is needed for a number of important reasons.[85]

In each space of the HyperCosmos a fundamental fermion has a set of creation and annihilation operators that undergo transformations when a General Relativistic transformation between a static and a non-static reference frame is made. The set of such transformations has a su($2^{r/2}$, $2^{r/2}$) symmetry group in the space with r space-time coordinates. The irreducible representation of the symmetry has $2^{r/2+2}$ real dimensions.

19.9 Dynamics and Unification

The unification of General Relativity and Quantum Theory is evident in the HyperCosmos at several levels. The complete set of symmetry groups, including the space-time group, is contained in each space. The dimension array with $2^{r/2+2} \times 2^{r/2+2} = 2^{r+4}$ real dimensions is the irreducible representation of a SU(2^{r+3}) symmetry group that is broken into the set of subgroups for internal symmetries and space-time for each HyperCosmos space. We view the SU(2^{r+3}) group as the initial group at the point of creation – the Big Bang, which has extremely large energy concentrated at one point.[86]

The explicit unity of the symmetry groups before separation into subgroups shows unification of the symmetry groups as one expects in Grand Unified Theories (GUTS) thus fulfilling one type of unification.

Another, perhaps deeper, form of unification stems from the CASe groups of GiFT. We remember the CASe groups were the source of the fundamental symmetry group dimension array. The CASe groups were defined based on the need for General Relativistic transformations of sets of creation/annihilation operators between static and non-static coordinate systems. General Relativity is thus at the heart of Quantum Theory. The requirement of covariance in both dynamic equations and in creation/annihilation operator expressions completes the unification program for Quantum Theory and General Relativity.

CASe is required by General Relativistic transformations of Quantum creation/annihilation operators to non-static coordinate systems. The dimension arrays of the symmetry groups follow from CASe and give an initial unified symmetry group for each HyperCosmos space. The Riemann-Christoffel curvature tensor embodies all the interactions including Gravitation and leads to a complete unified dynamics.. In the PseudoQuantum formulation its form is[87]

$$R'^{\beta}{}_{\sigma\nu\mu} = \Sigma \ \{R'_E{}^{1\beta}{}_{\sigma\nu\mu} + R'_E{}^{2\beta}{}_{\sigma\nu\mu} + R'_{SU(2)}{}^{1\beta}{}_{\sigma\nu\mu} + R'_{SU(2)}{}^{2\beta}{}_{\sigma\nu\mu} + R'_{DE}{}^{1\beta}{}_{\sigma\nu\mu} + R'_{DE}{}^{2\beta}{}_{\sigma\nu\mu} +$$

---

[84] Quantum Mechanics can be studied as an independent mathematical theory. Any ambiguities/issues that arise in that study must be resolvable physically since Nature has only one physical "answer" in any physical phenomena.

[85] The PseudoQuantum Theory is outlined in Appendix A.

[86] Blaha (2021d) shows the Big Bang and subsequent expansion can be viewed as analogous to vacuum polarization. It shows free field behavior at the Big Bang.

[87] See chapters 22ff of Blaha (2018e). This form is independently for each layer and for Normal and Dark sectors separately.

$$+ R'_{DSU(2)}{}^{1\beta}{}_{\sigma\nu\mu} + R'_{DSU(2)}{}^{2\beta}{}_{\sigma\nu\mu} + R'_{SU(3)}{}^{1\beta}{}_{\sigma\nu\mu} + R'_{SU(3)}{}^{2\beta}{}_{\sigma\nu\mu} + R'_{U}{}^{1\beta}{}_{\sigma\nu\mu} +$$

$$+ R'_{U}{}^{2\beta}{}_{\sigma\nu\mu} + R'_{V}{}^{1\beta}{}_{\sigma\nu\mu} + R'_{V}{}^{2\beta}{}_{\sigma\nu\mu} \} + R'_{B}{}^{1\beta}{}_{\sigma\nu\mu} + R'_{B}{}^{2\beta}{}_{\sigma\nu\mu} + R^{1\beta}{}_{\sigma\nu\mu} + R^{2\beta}{}_{\sigma\nu\mu}$$

where

$$R'_{E}{}^{1\beta}{}_{\sigma\nu\mu} = ig^{\beta}{}_{\sigma}F_{E}{}^{1}{}_{\nu\mu}$$
$$R'_{E}{}^{2\beta}{}_{\sigma\nu\mu} = ig^{\beta}{}_{\sigma}F_{DE}{}^{2}{}_{\nu\mu}$$

$$R'_{DE}{}^{1\beta}{}_{\sigma\nu\mu} = ig^{\beta}{}_{\sigma}F_{E}{}^{1}{}_{\nu\mu}$$
$$R'_{DE}{}^{2\beta}{}_{\sigma\nu\mu} = ig^{\beta}{}_{\sigma}F_{DE}{}^{2}{}_{\nu\mu}$$

$$R'_{SU(2)}{}^{1\beta}{}_{\sigma\nu\mu} = ig^{\beta}{}_{\sigma}F_{W}{}^{1}{}_{\nu\mu}$$
$$R'_{SU(2)}{}^{2\beta}{}_{\sigma\nu\mu} = ig^{\beta}{}_{\sigma}F_{DW}{}^{2}{}_{\nu\mu}$$

$$R'_{DSU(2)}{}^{1\beta}{}_{\sigma\nu\mu} = ig^{\beta}{}_{\sigma}F_{W}{}^{1}{}_{\nu\mu}$$
$$R'_{DSU(2)}{}^{2\beta}{}_{\sigma\nu\mu} = ig^{\beta}{}_{\sigma}F_{DW}{}^{2}{}_{\nu\mu}$$

$$R'_{SU(3)}{}^{1\beta}{}_{\sigma\nu\mu} = ig^{\beta}{}_{\sigma}F_{SU(3)}{}^{1}{}_{\nu\mu}$$
$$R'_{SU(3)}{}^{2\beta}{}_{\sigma\nu\mu} = ig^{\beta}{}_{\sigma}F_{SU(3)}{}^{2}{}_{\nu\mu}$$

$$R'_{U}{}^{1\beta}{}_{\sigma\nu\mu} = ig^{\beta}{}_{\sigma}F_{U}{}^{1}{}_{\nu\mu}$$
$$R'_{U}{}^{2\beta}{}_{\sigma\nu\mu} = ig^{\beta}{}_{\sigma}F_{U}{}^{2}{}_{\nu\mu}$$

$$R'_{V}{}^{1\beta}{}_{\sigma\nu\mu} = ig^{\beta}{}_{\sigma}F_{V}{}^{1}{}_{\nu\mu}$$
$$R'_{V}{}^{2\beta}{}_{\sigma\nu\mu} = ig^{\beta}{}_{\sigma}F_{V}{}^{2}{}_{\nu\mu}$$

$$R'_{B}{}^{1\beta}{}_{\sigma\nu\mu} = ig^{\beta}{}_{\sigma}B^{1}{}_{\nu\mu}$$
$$R'_{B}{}^{2\beta}{}_{\sigma\nu\mu} = ig^{\beta}{}_{\sigma}B^{2}{}_{\nu\mu}$$

The unification of Quantum Theory and General Relativity is completed by the combination of space-time dimensions and internal symmetry representation dimensions in a dimension array in each HyperCosmos space.

20. A Mathematical Veneer of Physics?

We have developed a theory of the Cosmos based on fermion creation and annihilation operators. These operators, while mathematically effective in describing particle creation and annihilation, must mask more complex Physical processes of creation and annihilation. Physically, particles do not simply pop into existence or disappear. There must be a reason for these operators' ability to handle complexities so simply.

The complexity of their simplicity is made more mysterious by the process of fractionating operators described earlier in Blaha (2022b) and here. An operator, fractionated to infinitesimal parts, still has a creation or annihilation property by eqs. 12.14, 12.18 and 12.19. Fractional parts also retain the internal symmetries of the whole.

So where are the "whirling gears or swirling currents" that Physics suggests should be implicit within creation or annihilation processes. "Where be their gambols?"

What we see in the mathematics is a mysterious abstraction. The Physics of these processes must be much deeper. It is encouraging that the limit of extreme fractionation yields commuting creation and annihilation operators—thus Classical Physics apparently. The particle dust within a fundamental particle should then be a classical fluid embodying a fluid dynamics.

A classical fluid should be governed by dynamic equations. For a non-interacting particle we expect that its dynamic equations are dependent on internal symmetry groups.[88] They should have a stationary solution—probably topological in nature. When interacting during creation or annihilation the structure of a particle should dynamically change with the output particle(s) entering into stationary states.

The nature of the fluid dynamics within particles remains to be understood—An enigma cloaked in the mystery of creation and annihilation.

---

[88] There is a suggestion that the inter-dust symmetry and interactions may be based on SU(4). See eq. 12.23 and surrounding comments.

Appendix A. Two-Tier and PseudoQuantum Theory Formalisms

The PseudoQuantum field Theory appears in Blaha (2021j) as well as.

S. Blaha, Phys. Rev. **D17**, 994 (1978).
S. Blaha, "The Local Definition of Asymptotic Particle States", IL Nuovo Cimento **49A**, 35 (1979).
S. Blaha, "New Framework for Gauge Field Theories", IL Nuovo Cimento **49A**, 113 (1979).

The PseudoQuantum Theory of Color Confinement appears in

S. Blaha, Phys. Rev. **D10**, 4268 (1974).
S. Blaha, Phys. Rev. D**11**, 2921 (1975).

PseudoQuantum Formalism

We are familiar with Quantum Theory, as it is usually developed, in our universe. In universes with higher space-time dimensions there is a need for a more expansive Quantum Field Theory. We presented this theory in earlier papers and books on Two-Tier Quantum Theory[89] and PseudoQuantum Field Theory.[90] Now it is GiFT. It resolved all divergences in ElectroWeak and Strong interaction theories. It also eliminated divergences in other types of Quantum Field Theories including theories with higher order derivatives and four fermion interactions.

In this book we focus on the higher space-time dimension spaces (universes) of the HyperCosmos. In these universes we find Two-Tier Quantum Theory is *absolutely* necessary to avoid divergences in perturbation theory calculations. The universes of the HyperCosmos have Feynman propagators and Perturbation Theory terms with integrations of the form:

$$\int d^n k\ f(k) \tag{A.1}$$

where n = 4, 8, 10, 12, 14, 16, and 18 (and also a less interesting case n = 2.)

In higher space-time dimensions Two-Tier quantization is essential. For example, in eight space-time dimensions the second order single fermion loop vacuum polarization is sextic divergent: $\int d^8k/k^2 \sim k^6$. Two-Tier Quantum Theory eliminates

[89] See Blaha (2005a) *Quantum Theory of the Third Kind.*
[90] Blaha (2018e) and earlier books.

divergences in Perturbation Theory with exponentiated Gaussian quadratic terms in momenta of the form:

$$\exp(-ak^2) \qquad (A.2)$$

where a is a constant. Two-Tier Quantum Theory is required for Perturbation Theory calculations in higher dimensions.

PseudoQuantum Field Theory is also needed in the HyperCosmos. It is needed for proper quantization in arbitrary coordinate systems that might be relevant in higher dimension HyperCosmos spaces. For example, consider coordinate systems for non-static space-times where no time-like Killing vector exists.[91]

It also enables canonical higher order derivative theories for quark confinement and other purposes. And it dovetails with Two-Tier Coordinates to "dress" bare fermion and boson particles.

A.1 Reasons For Two-Tier Quantum Theory

Originally Two-Tier coordinates were developed by this author to remove infinities that appear in perturbation theory calculations. We showed that the quantum smeared coordinates of Two-Tier Quantum Field Theory succeeded in removing all ultra-violet infinities in perturbation theory including the fermion triangle infinities.

Remarkably the high precision, "low" energy[92] predictions of QED remained true in Two-Tier QED and thus remained consistent with experiment to a hitherto unsurpassed level of accuracy. "Low" energy predictions in other quantum field theories also remained unchanged. At high energies, Two-Tier perturbation theory results are finite and consequently all ultra-violet infinities, to any order in perturbation theory, in *any number of space-time dimensions* were eliminated.

In addition to removing perturbation theory infinities, Two-Tier coordinates enable us to define finite theories of Quantum Gravity and 'non-renormalizable' quantum field theories based on polynomial Lagrangians, to tame vacuum fluctuations, to eliminate infinities associated with the Big Bang, and possibly to generate the explosive growth of the universe in its role as a type of Dark Energy.[93]

A.2 Two-Tier Features in 4-Dimensional Space-Time

Two-Tier Quantum Field Theory[94] is based on a new method[95] in the Calculus of Variations that uses two 'layers' of fields to introduce quantum coordinates. We shall consider this technique for the specific case of a massless vector field $Y^\mu(y)$ where the index μ ranges from 0 through 3. It is analogous to the electromagnetic field.

---

[91] B. DeWitt, Phys. Rep. **19**, 295 (1975) and references therein. S. Blaha, "New Framework for Gauge Field Theories", IL Nuovo Cimento **49A**, 113 (1979).

[92] Relative to a mass scale that was perhaps of the order of the Planck mass.

[93] See Blaha (2017b) and earlier books for details. This section is basically a summary of some features.

[94] See Blaha (2005a), and Blaha (2002), for discussions of this new method to eliminate infinities in quantum field theory calculations.

[95] See Blaha (2005a)..

The X^μ coordinate system, where it appears, has a c-number real part and a q-number imaginary part. Thus particle fields which are normally defined on real four-dimensional real space-time will now be defined on a "slightly" complex four-dimensional space-time:

$$X^\mu(y) = y^\mu + i\, Y^\mu(y)/M_c^2 \qquad (A.3)$$

where M_c is an extremely large mass of the order of the Planck mass or larger.

The $Y^\mu(y)$ field is a function of the space y coordinates. The real part of the space-time dimensions will be taken to be the space of real-valued y coordinates.[96]

The imaginary part of space-time coordinates is the massless $Y^\mu(y)$ vector quantum field that is suppressed by the very large mass scale. The effects of Quantum Dimensions only become appreciable in quantum field theory at energies of the order of M_c. At these energies exponential Gaussian factors in each particle (and ghost) propagator are generated by the Quantum Dimensions and serve to make *all* perturbation theory calculations ultra-violet finite – including calculations in Quantum Gravity. Later we will see that the Two-Tier formalism may be extended directly to the universes of the HyperCosmos with similar results – finiteness in Perturbation Theory.

The Two-Tier formalism introduces a new form of interaction that does not have the form of the simple polynomial interactions that have hitherto dominated quantum field theories. This form of interaction takes place via the composition of quantum fields and can be called a *Dimensional Interaction* or an *Interdimensional Interaction* since it affects particle behavior through Quantum Dimensions.

The basic *ansatz* of the Two-Tier formalism is to replace every appearance of a coordinate x in a quantum field with the variable

$$x^\mu \rightarrow X^\mu = (y^0, \mathbf{y} + \mathbf{Y}(y^0, \mathbf{y})/M_c^2) \qquad (A.4)$$

where $\mathbf{Y}(y^0, \mathbf{y})$ is the spatial part of a free massless vector field with features that are identical to the free QED field in the Radiation gauge.

Then one finds that the momentum space free field Feynman propagators G(k) of all particles acquires a Gaussian factor exp(h(k)):

$$G(k) \rightarrow G(k)\, \exp(h(k)) \qquad (A.5)$$

so that all perturbation theory diagrams are finite. The result is finite perturbative results for all calculations to any order in perturbation theory. Blaha (2005a) shows that Two-Tier theories are finite, Poincare covariant, and unitary. (See Blaha (2005a) for a complete discussion.)

[96] In a deeper theory the real part might also be a quantum field that undergoes a condensation to generate c-number coordinates. We will not consider this possibility in this book.

A.3 Two-Tier Quantum Coordinates Formalism

In this section we will introduce the basic Two-Tier formalism. Taking the Lagrangian described in Blaha (2005a):

$$\mathscr{L}(y) = \mathscr{L}_F\,(X^\mu(y))J + \mathscr{L}_C(X^\mu(y), \partial X^\mu(y)/\partial y^\nu, y) \qquad \text{(A.6)}$$

where

$$X^\mu(y) = y^\mu + i\,Y^\mu(y)/M_c^2 \qquad \text{(A.7)}$$

with M_C being a large mass scale, $Y_\mu(y)$ a vector quantum field, and where J is the absolute value of the Jacobian of the transformation from X to y coordinates:

$$J = |\partial(X)/\partial(y)|$$

The Lagrangian term $\mathscr{L}_C$ is

$$\mathscr{L}_C = +\tfrac{1}{4}\, M_c^4 F^{\mu\nu}F_{\mu\nu}$$

with

$$\begin{aligned} F_{\mu\nu} &= \partial X_\mu/\partial y^\nu - \partial X_\nu/\partial y^\mu \\ &\equiv i\,(\partial Y_\mu/\partial y^\nu - \partial Y_\nu/\partial y^\mu)/M_c^2 \end{aligned} \qquad \text{(A.8)}$$

The Lagrangian term $\mathscr{L}_F\,(X^\mu(y))$ contains the terms for scalar, fermion and other gauge terms in general. The sign in $\mathscr{L}_C$ is not negative – contrary to the conventional electromagnetic Lagrangian. The reason for this difference is that the quantum field part of X^μ is imaginary. Thus $\mathscr{L}_C$ ends up having the correct sign after taking account of the factor of i in the field strength $F_{\mu\nu}$.

Defining

$$F_{Y\mu\nu} = (\partial Y_\mu/\partial y^\nu - \partial Y_\nu/\partial y^\mu)$$

we see the Lagrangian assumes the form of the conventional electromagnetic Lagrangian:

$$\mathscr{L}_C = -\tfrac{1}{4}\, F_Y^{\mu\nu}F_{Y\mu\nu}$$

The action of this theory has the form

$$I = \int d^4y\, \mathscr{L}(y)$$

Since $X^\mu(y)$ has an imaginary part there would appear to be an issue with the conservation of probability (unitarity). *We show unitarity is not a problem later in this appendix in section A.8.*

A.4 Y^μ Gauge

The gauge invariance of the Lagrangian allows us to choose a convenient gauge. The gauge invariance of the full Lagrangian:

$$\mathscr{L}_S = L_F(\phi(X), \partial\phi/\partial X^\mu)\, J + \mathscr{L}_C(X^\mu(y), \partial X^\mu(y)/\partial y^\nu)$$

is based on the standard gauge invariance of $\mathscr{L}_C$, and the gauge invariance of $J\mathscr{L}_F$ in the form of translational invariance

$$X^\mu(y) \rightarrow X^\mu(y) + \delta X^\mu(y)$$

for the special case of a translation of X with the form of a gauge transformation:

$$\delta X^\mu(y) = \partial\Lambda(y)/\partial y_\mu$$

In this case we find

$$\int d^4y\ \Lambda(y)\, \partial\,[\,J\,\partial/\partial X^\mu\, \mathscr{F}_{F\mu\nu}\,]/\partial y_\nu = 0 \tag{A.9}$$

after a partial integration. Thus we have the differential conservation law:

$$\partial\,[\,J\,\partial\mathscr{F}_{F\mu\nu}/\partial X^\mu]/\partial y_\nu = 0$$

since $\Lambda(y)$ is arbitrary. This conservation law is trivially obeyed:

$$\partial\mathscr{F}_{F\mu\nu}/\partial X^\mu = 0 \tag{A.10}$$

Thus translational invariance in the $\mathscr{L}_F$ sector together with standard gauge invariance in the $\mathscr{L}_C$ sector automatically guarantees Y field gauge invariance of the total Lagrangian. We use the separate invariance of each term of

$$L = \int d^4y\ [\mathscr{L}_F\, J + \mathscr{L}_C\,] = \int d^4X\ \mathscr{L}_F + \int d^4y\ \mathscr{L}_C = L_F + L_C$$

under a constant translation $X^\mu \rightarrow X^\mu + \delta X^\mu$ where δX^μ is constant. Then we consider a position dependent translation/gauge transformation, which taken together with the above equation, establishes the invariance under the position dependent translation/gauge transformation.

An alternate approach that leads to the same result is to start with the particle part of the Lagrangian $\mathscr{L}_F$ rewritten to be invariant under general coordinate transformations, as it must, when we generalize to include General Relativity. Since position dependent translations are a form of general coordinate transformation the full theory must be invariant under position dependent translations due to invariance under general coordinate transformations.

Having established invariance under gauge transformations we now choose to use the most convenient gauge – the radiation gauge[97]:

$$\partial Y^i/\partial y^i = 0 \tag{A.11}$$

where i = 1, 2, 3. In the absence of external sources, we set

$$Y^0 = 0$$

since Y^0 does not have a canonically conjugate momentum. A conventional treatment leads to the equal time commutation relations:

$$\begin{gathered}[Y^\mu(\mathbf{y}, y^0), Y^\nu(\mathbf{y}', y^0)] = [\pi^\mu(\mathbf{y}, y^0), \pi^\nu(\mathbf{y}', y^0)] = 0 \\ [\pi_j(\mathbf{y}, y^0), Y_k(\mathbf{y}', y^0)] = -i\,\delta^{tr}{}_{jk}(\mathbf{y} - \mathbf{y}')\end{gathered} \tag{A.12}$$

where (Note the locations of the j indexes above introduce a minus sign.)

$$\begin{gathered}\pi^k = \partial\mathscr{L}_C/\partial Y_k' \\ \pi^0 = 0\end{gathered}$$

$$\delta^{tr}{}_{jk}(\mathbf{y} - \mathbf{y}') = \int d^3k\, e^{i\,\mathbf{k}\cdot(\mathbf{y} - \mathbf{y}')}(\delta_{jk} - k_j k_k/\mathbf{k}^2)/(2\pi)^3 \tag{A.13}$$

$$Y_k' = \partial Y_k/\partial y^0$$

The Radiation gauge reveals the two degrees of freedom that are present in the vector potential. The Fourier expansion of the vector potential is:

$$Y^i(y) = \int d^3k\, N_0(k) \sum_{\lambda=1}^{2} \varepsilon^i(k, \lambda)[a(k,\lambda)\, e^{-ik\cdot y} + a^\dagger(k,\lambda)\, e^{ik\cdot y}] \tag{A.14}$$

where

$$N_0(k) = [(2\pi)^3 2\omega_k]^{-1/2}$$

and (since m = 0)

$$\omega_k = (\mathbf{k}^2)^{1/2} = k^0$$

with $\vec{\epsilon}(k, \lambda)$ being the polarization unit vectors for $\lambda = 1{,}2$ and $k^\mu k_\mu = 0$.

---

[97] It is also possible to quantize using an indefinite metric that preserves manifest Lorentz covariance as was done by Gupta and Bleuler for the electromagnetic field. We will use the Gupta-Bleuler approach later to establish covariance under special relativity later. Now we opt for manifest positivity and use the radiation gauge.

The further development of Two-Tier theory is described in Part 3 of Blaha (2005a).

A.5 Two-Tier Uncertainty Principle

The Uncertainty Relation for Quantum Mechanics is based on the coordinate-momentum commutator. Similarly, in defining Quantum Coordinates we have established a commutation relation based on Y^μ. Its conjugate momentum is

$$P^\mu(y) = i\pi_Y{}^\mu(y)/M_c{}^2 \tag{A.15}$$

where

$$\pi_Y{}^\mu(y) = -\, dY^\mu/(y)dy^0$$

In the Radiation gauge (eq. A.11) we see

$$\pi_Y{}^0(y) = 0 \tag{A.16}$$

and

$$[X^0, P^0] = [y^0, p^0] = 0 \tag{A.17}$$

The non-zero equal time commutation relation expressing a quantum Uncertainty Relation is

$$[P^i(\mathbf{y}, y^0), Y^k(\mathbf{y}', y^0)] = -[\pi_Y{}^i(y), Y^k(y),]/M_c{}^4 \tag{A.18}$$

$$= i\delta^{trik}(\mathbf{y} - \mathbf{y}')/M_c{}^4$$

using eq. A.12.

At low energy the impact of the Two-Tier Uncertainty Relation is diminished (more or less eliminated) by the factor $M_c{}^4$. Then Two-Tier Quantum Field Theory becomes ordinary Quantum Field Theory with the same results in Perturbation Theory. This limit is described in detail in Blaha (2005a), which appears in in Appendix B.

In the "low" energy limit the conventional Heisenberg Uncertainty Condition becomes evident as shown by Heitler (1954) and others. *Conventional Quantum Mechanics is a result of Second Quantization.*

A.6 Quantum Mechanics vs. Quantum Field Theory vs. Two-Tier Quantum Theory

Historically, Quantum Mechanics predated Quantum Field Theory, which, in turn, predates the Two-Tier Quantum Theory developed by the author in the early 2000's.

Logically, Two-Tier Quantum Theory, which embodies an Uncertainty Relation at ultrahigh energies, is the fundamental Quantum Theory. It is the predecessor of Quantum Field Theory, which only embodies the Heisenberg Uncertainty Relation. Quantum Field Theory is flawed by high energy divergences just as atomic physics, before the Bohr atom and Quantum Mechanics, was flawed by infinities in hydrogen atom models.

Quantum Field Theory (a "low" energy theory) implies Quantum Mechanics[98] as shown in Heitler[99] as well as elsewhere.

Quantum Mechanics is often treated as a complete, self-contained theory. On occasion paradoxes and ambiguities are found in Quantum Mechanics studies. They may be resolved within Quantum Mechanics. Any, that are not so resolved, should be considered within the framework of Quantum Field Theory, which is the ultimate forum for all quantum phenomena at "low" energy.

Note the trend of quantum theories, from the earliest at "low" energy theory, Quantum Mechanics, to the highest energy quantum theory: Two-Tier Quantum Theory,

Two-Tier Quantum Field Theory is a welcome extension of Quantum Theory to the deepest levels of Quantum Theory.

A.7 Two-Tier Perturbation Theory

The form of Two-Tier Perturbation Theory is similar to the Perturbation Theory of conventional Quantum Field Theory. It is described in *Quantum Theory of the Third Kind.* The reduction development for the U-matrix and the S-matrix are presented there.

A.8 Two-Tier Unitarity

The unitarity of Two-Tier Quantum Field Theory can be viewed in the cases of "low" energy phenomena and "high" energy phenomena with the scale mass of M_c separating the cases.

In the case of "low" energy phenomena the Two-Tier S-matrix gives results identical to conventional S-matrix results. Thus there is no conflict with unitarity for Two-Tier S-matrix results at "low" energy.

At "high" energy there are two potential issues: asymptotic states containing Y quanta; and conservation of probability. The first issue is resolved in the chapter 6 discussion in Appendix B of Blaha (2021i).

The second issue leads to a requirement to renormalize S-marix probability amplitudes such that their absolute values squared sum to unity – the unitarity condition. If we let S-matrix elements have the form S_{fi} where i represents an initial state and f represents one of its final states, then

$$\sum_n S_{nf}{}^* S_{ni} = \delta_{fi} \qquad \text{(A.19)}$$

expresses the conventional unitarity condition. In Two-Tier Quantum Field Theory the unitarity condition for energies much less than M_c is the same as eq. A.19 since perturbation theory results are the same as conventional Quantum Field Theory. The

---

[98] Quantum Mechanics is often treated as an independent theory. This practice may be valid mathematically but it is not valid physically. For example, many phenomena in atomic theory require explanation in terms of Quantum Field Theory.

[99] Heitler (1954).

sum over intermediate states n is restricted to states whose total energy is less than M_c by energy conservation.

For initial states i with energy E_i of the order of or greater than M_c, the intermediate state total energy and the final state total energy are also E_i by energy conservation. In this case the sum in eq. A.19 *may* be

$$\sum_n S_{nf}{}^* S_{ni} = g_i \delta_{fi} \tag{A.20}$$

where g_i is a constant term dependent on the state i. In this circumstance unitarity may be recovered by redefining S_{ni} and S_{nf} as

$$S_{ni}' = S_{ni}/\surd g_i \tag{A.21}$$
$$S_{nf}' = S_{nf}/\surd g_f \tag{A.22}$$

Then the *normalized* Two-Tier unitarity condition can be expressed as

$$\sum_n S_{nf}'^* S_{ni}' = \delta_{fi} \tag{A.23}$$

with the understanding that $g_i = 1$ for initial states with energies much less than M_c.

Thus Two-Tier Quantum Field Theory satisfies an enhanced unitarity condition.

A.8.1 Two-Tier Analyticity

Two-Tier Quantum Theory has an apparent issue with analyticity. It is resolved by noting that lower energy perturbation theory results in Two-Tier calculations are the same as conventional quantum theory although the high energy limits are not divergent in Two-Tier calculations.

At high energy Two-Tier results differ from conventional results. This circumstance might be an issue but there are no experimental results on analyticity at high energy. So the point is mute—especially considering that HyperCosmos perturbation calculations for large space-time dimensions can currently only be done using Two-Tier quantum theory.

A.9 Two-Tier Quantum Space Theory

Quantum Space Theory is the theory of particles containing spaces internally. It was recently developed by the author.[100] This theory can directly put in the form of Two-Tier Quantum Space Theory by following the procedure described in section A.2. The structure of the Octonion Spaces Spectrum is not changed by this formulation.

A.10 CQ Mechanics – A Union of Classical and Quantum Mechanics

The author developed a larger theory, CQ Mechanics, containing both Quantum Mechanics and Classical Mechanics that is described in Blaha (2016f). In this theory a phenomenon can be described as classical in one limit and as quantum in another limit.

---

[100] Blaha (2021f) and (2021g).

The description is determined by an angle that specifies one, or the other, limiting case. Intermediate cases combining both quantum and classical are also present.

This theory can bridge the classical and quantum regimes. We discuss some applications in Blaha (20016f): a generalized Feynman path integral formalism, a generalized Schrödinger equation, a generalized Boltzmann equation, the Fokker-Planck equation, a generalized approach to quantum and classical chaos, and to quantum entanglement as well as semi-quantum entanglement. Our formalism applies to Quantum Mechanics as well as the path integrals, the Fokker-Planck equation and the Boltzmann equation.

This "Mechanics" theory has an analogue at the field theoretic level, PseudoQuantum Field Theory that is briefly described next.

A.11 PseudoQuantum Field Theory

PseudoQuantum Field Theory[101] was developed by the author in the 1970's and presented in a series of articles including the papers below.

S. Blaha, "The Local Definition of Asymptotic Particle States", IL Nuovo Cimento **49A**, 35 (1979). It describes the PseudoQuantization of boson and fermion field theories for use in the quantization of fields in universes and the Megaverse.

S. Blaha, "New Framework for Gauge Field Theories", IL Nuovo Cimento **49A**, 113 (1979). It describes the PseudoQuantization of gauge field theories for the purposes of defining higher derivative field theories and for use in the quantization of fields in universes and the Megaverse.

PseudoQuantum Field Theory (and its Quantum Mechanics analogue CQ Mechanics[102]) originated in the need to second quantize in unusual coordinate systems, and in curved space-time coordinate systems. It is also very relevant for the canonical formulation of higher derivative field theories for quark confinement and other applications.

PseudoQuantum Field Theory is formulated by duplicating all fields, both fermion and boson fields, in a "normal" Lagrangian theory. Scalar field theory provides a simple example that illustrates the PseudoQuantum Field Theory procedure.

We duplicate a scalar field generating two scalar fields: $\varphi_1(x)$ and $\varphi_2(x)$. We choose $\varphi_1(x)$ to have a zero equal time commutator with $d\varphi_1(x)/dx^0$ and $\varphi_2(x)$ to have a conventional equal time commutator with $d\varphi_2(x)/dx^0$. Conceptually $\varphi_1(x)$ is a "classical" field and $\varphi_2(x)$ is a quantum field. A Lagrangian that implements these choices of commutation relations is:

$$\mathcal{L} = \partial^\mu \varphi_1(x)\partial_\mu\varphi_2(x) - \tfrac{1}{2}\,\partial^\mu \varphi_1(x)\partial_\mu\varphi_1(x) - m_2^{\,2}\, \varphi_1(x)\varphi_2(x) + \tfrac{1}{2}\, m_1^{\,2}\, \varphi_1(x)^2 \quad \text{(A.24)}$$

---

[101] Appendix C describes PseudoQuantum Theory features in some detail.

[102] See Blaha (2016f) for details, which contains Blaha (2016f). CQ Mechanics encompasses both classical mechanics and quantum mechanics, and provides a method of rotating between them. It has applications to transitions between Quantum/Semi-Classical Entanglement, and Quantum/Classical Path Integrals, and Quantum/Classical Chaos.

$$(\Box + m_2^2)\varphi_1(x) = 0 \qquad (A.25)$$

$$(\Box + m_2^2)\varphi_2(x) - (\Box + m_1^2)\varphi_1(x) = 0 \qquad (A.26)$$

The canonical momenta are

$$\pi_1 = d\varphi_2(x)/dt - d\varphi_1(x)/dt \qquad (A.27)$$
$$\pi_2 = d\varphi_1(x)/dt \qquad (A.28)$$

and the equal time commutation relations are

$$[\varphi_i(x), \pi_j(y)] = i\delta_{ij}\delta(\mathbf{x} - \mathbf{y}) \qquad (A.29)$$
$$[\varphi_i(x), \varphi_j(y)] = [\pi_i(x), \pi_j(y)] = 0 \qquad (A.30)$$

implying

$$[\varphi_1(x), d\varphi_1(y)/dt] = 0 \qquad (A.31)$$
$$[\varphi_2(x), d\varphi_2(y)/dt] = i\delta(\mathbf{x} - \mathbf{y}) \qquad (A.32)$$
$$[\varphi_1(x), d\varphi_2(y)/dt] = i\delta(\mathbf{x} - \mathbf{y}) \qquad (A.33)$$

A.11.1 Color Confinement

Appendix F contains a paper on a non-Abelian gauge field theory that gives explicit color confinement using a PseudoQuantum formulation that leads to a confining r-potential.

If we set $m_2 = 0$ in the above example, we see a fourth order field equation results

$$\Box^2\varphi_2(x) = 0 \qquad (A.34)$$

Color confinement results in the non-Abelian gauge field case in a similar manner.

A.11.2 PseudoQuantum Quantization for Non-Static Coordinate Systems

The PseudoQuantum formalism was developed by the author for "normal" and non-static coordinate systems in the 1970s.

A.11.3 Advantages of PseudoQuantum Quantization

In this section we point out some of its advantages in a variety of field theory contexts that are relevant for the HyperCosmos and Quantum Field Theory in general.

Some advantages of PseudoQuantum Field Theory are:

4. Quantization in any coordinate system in flat or curved space-times with an invariant definition of asymptotic particle states. *This is especially important for the higher dimension spaces of the HyperCosmos.* An n particle asymptotic state in one coordinate system is a unitarily equivalent n particle asymptotic state in any other coordinate system. Therefore particle number is invariant under change of coordinate system. This is important for the Unified SuperStandard Theory in curved space-times. It is also important for quantization in higher dimensional spaces such as those of the HyperCosmos. The method was

developed in the late 1970's by the author to provide a quantization procedure which supports a unique particle interpretation of states in arbitrary non-static space-times where no global time-like coordinate (Killing vector) exists. PseudoQuantum Field Theory which we developed in a series of books[103] also can be formulated in the Octonion Spectrum of spaces. For example, we can use it to implement the Higgs Mechanism to generate particle masses and symmetry breaking.

5. PseudoQuantum Field Theory enables one to define Higgs particle dynamics in such a way that a non-zero vacuum expectation value cleanly separates from the quantum field part of the Higgs fields. This technique can be used in symmetry breaking mechanisms, mass generation, and possible generation of coupling constants as vacuum expectation values.

6. It supports the canonical definition of higher derivative field theories through the use of the Ostrogradski bootstrap. We have used it to construct a fourth order theory of the Strong interaction that has color confinement and a linear r potential. The potential part of this theory was used by the Cornell group to calculate the Charmonium spectrum. (See Blaha (2017b) for details.)

A.12 Combination of Two-Tier and PseudoQuantum Formalisms

These formalisms can be directly combined with no issues.

A.13 A Model Illustrating Scalar Field Quantization Using X^μ

We begin by considering the case of a scalar quantum field theory. We assume a real underlying y subspace. Since X^μ is a set of coordinates, we choose to define a scalar field ϕ as a function of X^μ, which, in turn, is a function of the y^ν coordinates. We will provisionally second quantize ϕ treating X^μ as c-number coordinates using a conventional approach.[104]

We assume a Lagrangian, with the momentum conjugate to ϕ:

$$\pi_\phi = \partial \mathrm{L_F} / \partial \phi' \equiv \partial \mathrm{L_F} / \partial(\partial \phi / \partial X^0) \qquad \text{(A.35)}$$

Following the canonical quantization procedure, π and ϕ become Hermitian operators with equal time ($X^0 = X^{0\prime}$) commutation rules:

---

[103] See Blaha (2017b) for the discussion of the PseudoQuantum field theory formalism for Higgs particles in our Extended Standard Model. See chapter 20 of Blaha (2017b), and earlier books, for a more detailed view than that presented here.

[104] Some texts are: Bogoliubov, N. N., Shirkov, D. V., *Introduction to the Theory of Quantized Fields* (Wiley-Interscience Publishers Inc., New York, 1959); Bjorken, J. D., Drell, S. D., *Relativistic Quantum Fields* (McGraw-Hill, New York, 1965); Huang, K., *Quarks, Leptons & Gauge Fields Second Edition* (World Scientific, River Edge, NJ, 1992); Kaku, M., *Quantum Field Theory* (Oxford University Press, New York, 1993); Weinberg, S., *The Quantum Theory of Fields* (Cambridge University Press, New York, 1995).

$$[\phi(X), \phi(X')] = [\pi_\phi(X), \pi_\phi(X')] = 0 \qquad (A.36)$$
$$[\pi_\phi(X), \phi(X')] = -i\,\delta^3(\mathbf{X} - \mathbf{X'})$$

The standard Fourier expansion of the solution to the Klein-Gordon equation is:

$$\phi(X) = \int d^3p\, N_m(p)\, [a(p)\, e^{-ip\cdot X} + a^\dagger(p)\, e^{ip\cdot X}] \qquad (A.36a)$$
$$= \int d^{3k}p\, N_m(p)\, [a(p) \exp(-ip\cdot(y + Y/M_c^2)) +$$
$$+ a^\dagger(p) \exp(ip\cdot(y + Y/M_c^2))] \qquad A.36b)$$

where

$$N_m(p) = [(2\pi)^3 2\omega_p]^{-1/2}$$

and

$$\omega_p = (\mathbf{p}^2 + m^2)^{1/2}$$

The commutation relations of the Fourier coefficient operators are:

$$[a(p), a^\dagger(p')] = \delta^3(\mathbf{p} - \mathbf{p'})$$
$$[a^\dagger(p), a^\dagger(p')] = [a(p), a(p')] = 0$$

The reader will recognize the quantization procedure is formally identical to the standard canonical quantization procedure of a free scalar quantum field.

In the case of spin ½, spin 1 and spin 2 fields the standard quantization procedure *in terms of the X coordinate system* can also be followed in a way similar to the procedure in standard texts.

A.14 Scalar Feynman Propagators

The momentum space free field Feynman propagators G...(k) of all particles and ghosts in all Two-Tier Quantum Field Theories acquires a Gaussian factor exp(h(k)):

$$G...(k) \to G...(k) \exp(h(k))$$

so that all perturbation theory diagrams are finite. The result is a finite perturbation in all calculations to any order in perturbation theory. Blaha (2005a) shows that Two-Tier theories are finite, Poincare covariant, and unitary.

REFERENCES

Akhiezer, N. I., Frink, A. H. (tr), 1962, *The Calculus of Variations* (Blaisdell Publishing, New York, 1962).

Bjorken, J. D., Drell, S. D., 1964, *Relativistic Quantum Mechanics* (McGraw-Hill, New York, 1965).

Bjorken, J. D., Drell, S. D., 1965, *Relativistic Quantum Fields* (McGraw-Hill, New York, 1965).

Blaha, S., 1995, *C++ for Professional Programming* (International Thomson Publishing, Boston, 1995).

_______, 1998, *Cosmos and Consciousness* (Pingree-Hill Publishing, Auburn, NH, 1998 and 2002).

_______, 2002, *A Finite Unified Quantum Field Theory of the Elementary Particle Standard Model and Quantum Gravity Based on New Quantum Dimensions™ & a New Paradigm in the Calculus of Variations* (Pingree-Hill Publishing, Auburn, NH, 2002).

_______, 2004, *Quantum Big Bang Cosmology: Complex Space-time General Relativity, Quantum Coordinates™ Dodecahedral Universe, Inflation, and New Spin 0, ½, 1 & 2 Tachyons & Imagyons* (Pingree-Hill Publishing, Auburn, NH, 2004).

______, *2005a, Quantum Theory of the Third Kind: A New Type of Divergence-free Quantum Field Theory Supporting a Unified Standard Model of Elementary Particles and Quantum Gravity based on a New Method in the Calculus of Variations* (Pingree-Hill Publishing, Auburn, NH, 2005).

______, 2005b, *The Metatheory of Physics Theories, and the Theory of Everything as a Quantum Computer Language* (Pingree-Hill Publishing, Auburn, NH, 2005).

______, 2005c, *The Equivalence of Elementary Particle Theories and Computer Languages: Quantum Computers, Turing Machines, Standard Model, Superstring Theory, and a Proof that Gödel's Theorem Implies Nature Must Be Quantum* (Pingree-Hill Publishing, Auburn, NH, 2005).

______, 2006a, *The Foundation of the Forces of Nature* (Pingree-Hill Publishing, Auburn, NH, 2006).

______, 2006b, *A Derivation of ElectroWeak Theory based on an Extension of Special Relativity; Black Hole Tachyons; & Tachyons of Any Spin.* (Pingree-Hill Publishing, Auburn, NH, 2006).

_______, 2007a, *Physics Beyond the Light Barrier: The Source of Parity Violation, Tachyons, and A Derivation of Standard Model Features* (Pingree-Hill Publishing, Auburn, NH, 2007).

______, 2007b, *The Origin of the Standard Model: The Genesis of Four Quark and Lepton Species, Parity Violation, the ElectroWeak Sector, Color SU(3), Three Visible Generations of Fermions, and One Generation of Dark Matter with Dark Energy* (Pingree-Hill Publishing, Auburn, NH, 2007).

______, *2008a, A Direct Derivation of the Form of the Standard Model From GL(16) (Pingree-Hill Publishing, Auburn, NH, 2008).*

______, 2008b, *A Complete Derivation of the Form of the Standard Model With a New Method to Generate Particle Masses Second Edition* (Pingree-Hill Publishing, Auburn, NH, 2008)

______, 2009, *The Algebra of Thought & Reality: The Mathematical Basis for Plato's Theory of Ideas, and Reality Extended to Include A Priori Observers and Space-Time Second Edition* (Pingree-Hill Publishing, Auburn, NH, 2009).

______, 2010a, *Operator Metaphysics: A New Metaphysics Based on a New Operator Logic and a New Quantum Operator Logic that Lead to a Mathematical Basis for Plato's Theory of Ideas and Reality* (Pingree-Hill Publishing, Auburn, NH, 2010).

______, 2010b, *The Standard Model's Form Derived from Operator Logic, Superluminal Transformations and GL(16)* (Pingree-Hill Publishing, Auburn, NH, 2010).

______, 2010c, *SuperCivilizations: Civilizations as Superorganisms* (McMann-Fisher Publishing, Auburn, NH, 2010).

______, 2011a, *21st Century Natural Philosophy Of Ultimate Physical Reality* (McMann-Fisher Publishing, Auburn, NH, 2011).

______, 2011b, *All the Universe! Faster Than Light Tachyon Quark Starships & Particle Accelerators with the LHC as a Prototype Starship Drive Scientific Edition* (Pingree-Hill Publishing, Auburn, NH, 2011).

______, 2011c, *From Asynchronous Logic to The Standard Model to Superflight to the Stars* (Blaha Research, Auburn, NH, 2011).

______, 2012a, *From Asynchronous Logic to The Standard Model to Superflight to the Stars volume 2: Superluminal CP and CPT, U(4) Complex General Relativity and The Standard Model, Complex Vierbein General Relativity, Kinetic Theory, Thermodynamics* (Blaha Research, Auburn, NH, 2012).

______, 2012b, *Standard Model Symmetries, And Four And Sixteen Dimension Complex Relativity; The Origin Of Higgs Mass Terms* (Blaha Reasearch, Auburn, NH, 2012).

______, 2013a, *Multi-Stage Space Guns, Micro-Pulse Nuclear Rockets, and Faster-Than-Light Quark-Gluon Ion Drive Starships* (Blaha Research, Auburn, NH, 2013).

______, 2013b, *The Bridge to Dark Matter; A New Sibling Universe; Dark Energy; Inflatons; Quantum Big Bang; Superluminal Physics; An Extended Standard Model Based on Geometry* (Blaha Reasearch, Auburn, NH, 2013).

______, 2014a, *Universes and Megaverses: From a New Standard Model to a Physical Megaverse; The Big Bang; Our Sibling Universe's Wormhole; Origin of the Cosmological Constant, Spatial Asymmetry of the Universe, and its Web of Galaxies; A Baryonic Field between Universes and Particles; Megaverse Extended Wheeler-DeWitt Equation* (Blaha Reasearch, Auburn, NH, 2014).

______, 2014b, *All the Megaverse! Starships Exploring the Endless Universes of the Cosmos Using the Baryonic Force* (Blaha Research, Auburn, NH, 2014).

______, 2014c, *All the Megaverse! II Between Megaverse Universes: Quantum Entanglement Explained by the Megaverse Coherent Baryonic Radiation Devices – PHASERs Neutron Star Megaverse Slingshot Dynamics Spiritual and UFO Events, and the Megaverse Microscopic Entry into the Megaverse* (Blaha Research, Auburn, NH, 2014).

______, 2015a, *PHYSICS IS LOGIC PAINTED ON THE VOID: Origin of Bare Masses and The Standard Model in Logic, U(4) Origin of the Generations, Normal and Dark Baryonic Forces, Dark Matter, Dark Energy, The Big Bang, Complex General Relativity, A Megaverse of Universe Particles* (Blaha Research, Auburn, NH, 2015).

______, 2015b, *PHYSICS IS LOGIC Part II: The Theory of Everything, The Megaverse Theory of Everything, U(4)⊗U(4) Grand Unified Theory (GUT), Inertial Mass = Gravitational Mass, Unified Extended Standard Model and a New Complex General Relativity with Higgs Particles, Generation Group Higgs Particles* (Blaha Research, Auburn, NH, 2015).

______, 2015c, *The Origin of Higgs ("God") Particles and the Higgs Mechanism: Physics is Logic III, Beyond Higgs – A Revamped Theory With a Local Arrow of Time, The Theory of Everything Enhanced, Why Inertial Frames are Special, Universes of the Mind* (Blaha Research, Auburn, NH, 2015).

______, 2015d, *The Origin of the Eight Coupling Constants of The Theory of Everything: U(8) Grand Unified Theory of Everything (GUTE), S^8 Coupling Constant Symmetry, Space-Time Dependent Coupling Constants, Big Bang Vacuum Coupling Constants, Physics is Logic IV* (Blaha Research, Auburn, NH, 2015).

______, 2016a, *New Types of Dark Matter, Big Bang Equipartition, and A New U(4) Symmetry in the Theory of Everything: Equipartition Principle for Fermions, Matter is 83.33% Dark, Penetrating the Veil of the Big Bang, Explicit QFT Quark Confinement and Charmonium, Physics is Logic V* (Blaha Research, Auburn, NH, 2016).

______, 2016b, *The Periodic Table of the 192 Quarks and Leptons in The Theory of Everything: The U(4) Layer Group, Physics is Logic VI* (Blaha Research, Auburn, NH, 2016).

______, 2016c, *New Boson Quantum Field Theory, Dark Matter Dynamics, Dark Matter Fermion Layer Mixing, Genesis of Higgs Particles, New Layer Higgs Masses, Higgs Coupling Constants, Non-Abelian Higgs Gauge Fields, Physics is Logic VII* (Blaha Research, Auburn, NH, 2016).

______, 2016d, *Unification of the Strong Interactions and Gravitation: Quark Confinement Linked to Modified Short-Distance Gravity; Physics is Logic VIII* (Blaha Research, Auburn, NH, 2016).

______, 2016e, *MoND: Unification of the Strong Interactions and Gravitation II, Quark Confinement Linked to Large-Scale Gravity, Physics is Logic IX* (Blaha Research, Auburn, NH, 2016).

______, 2016f, *CQ Mechanics: A Unification of Quantum & Classical Mechanics, Quantum/Semi-Classical Entanglement, Quantum/Classical Path Integrals, Quantum/Classical Chaos* (Blaha Research, Auburn, NH, 2016).

______, 2016g, *GEMS Unified Gravity, ElectroMagnetic and Strong Interactions: Manifest Quark Confinement, A Solution for the Proton Spin Puzzle, Modified Gravity on the Galactic Scale* (Pingree Hill Publishing, Auburn, NH, 2016).

______, 2016h, *Unification of the Seven Boson Interactions based on the Riemann-Christoffel Curvature Tensor* (Pingree Hill Publishing, Auburn, NH, 2016).

______, 2017a, *Unification of the Eleven Boson Interactions based on 'Rotations of Interactions'* (Pingree Hill Publishing, Auburn, NH, 2017).

______, 2017b, *The Origin of Fermions and Bosons, and Their Unification* (Pingree Hill Publishing, Auburn, NH, 2017).

______, 2017c, *Megaverse: The Universe of Universes* (Pingree Hill Publishing, Auburn, NH, 2017).

______, 2017d, *SuperSymmetry and the Unified SuperStandard Model* (Pingree Hill Publishing, Auburn, NH, 2017).

______, 2017e, *From Qubits to the Unified SuperStandard Model with Embedded SuperStrings: A Derivation* (Pingree Hill Publishing, Auburn, NH, 2017).

______, 2017f, *The Unified SuperStandard Model in Our Universe and the Megaverse: Quarks, ... ,* (Pingree Hill Publishing, Auburn, NH, 2017).

______, 2018a, *The Unified SuperStandard Model and the Megaverse SECOND EDITION A Deeper Theory based on a New Particle Functional Space that Explicates Quantum Entanglement Spookiness (Volume 1)* (Pingree Hill Publishing, Auburn, NH, 2018).

______, 2018b, *Cosmos Creation: The Unified SuperStandard Model, Volume 2, SECOND EDITION* (Pingree Hill Publishing, Auburn, NH, 2018).

______, 2018c, *God Theory (*Pingree Hill Publishing, Auburn, NH, 2018).

______, 2018d, *Immortal Eye: God Theory: Second Edition* (Pingree Hill Publishing, Auburn, NH, 2018).

______, 2018e, *Unification of God Theory and Unified SuperStandard Model THIRD EDITION* (Pingree Hill Publishing, Auburn, NH, 2018).

______, 2019a, *Calculation of: QED α = 1/137, and Other Coupling Constants of the Unified SuperStandard Theory* (Pingree Hill Publishing, Auburn, NH, 2019).

______, 2019b, *Coupling Constants of the Unified SuperStandard Theory SECOND EDITION* (Pingree Hill Publishing, Auburn, NH, 2019).

______, 2019c, *New Hybrid Quantum Big_Bang–Megaverse_Driven Universe with a Finite Big Bang and an Increasing Hubble Constant* (Pingree Hill Publishing, Auburn, NH, 2019).

______, 2019d, *The Universe, The Electron and The Vacuum* (Pingree Hill Publishing, Auburn, NH, 2019).

______, 2019e, *Quantum Big Bang – Quantum Vacuum Universes (Particles)* (Pingree Hill Publishing, Auburn, NH, 2019).

______, 2019f, *The Exact QED Calculation of the Fine Structure Constant Implies ALL 4D Universes have the Same Physics/Life Prospects* (Pingree Hill Publishing, Auburn, NH, 2019).

______, 2019g, *Unified SuperStandard Theory and the SuperUniverse Model: The Foundation of Science* (Pingree Hill Publishing, Auburn, NH, 2019).

______, 2020a, *Quaternion Unified SuperStandard Theory (The QUeST) and Megaverse Octonion SuperStandard Theory (MOST)* (Pingree Hill Publishing, Auburn, NH, 2020).

______, 2020b, *United Universes Quaternion Universe - Octonion Megaverse* (Pingree Hill Publishing, Auburn, NH, 2020).

______, 2020c, *Unified SuperStandard Theories for Quaternion Universes & The Octonion Megaverse* (Pingree Hill Publishing, Auburn, NH, 2020).

______, 2020d, *The Essence of Eternity: Quaternion & Octonion SuperStandard Theories* (Pingree Hill Publishing, Auburn, NH, 2020).

______, 2020e, *The Essence of Eternity II* (Pingree Hill Publishing, Auburn, NH, 2020).

______, 2020f, *A Very Conscious Universe* (Pingree Hill Publishing, Auburn, NH, 2020).

______, 2020g, *Hypercomplex Universe* (Pingree Hill Publishing, Auburn, NH, 2020).

______, 2020h, *Beneath the Quaternion Universe* (Pingree Hill Publishing, Auburn, NH, 2020).

______, 2020i, *Why is the Universe Real? From Quaternion & Octonion to Real Coordinates* (Pingree Hill Publishing, Auburn, NH, 2020).

______, 2020j, *The Origin of Universes: of Quaternion Unified SuperStandard Theory (QUeST); and of the Octonion Megaverse (UTMOST)* (Pingree Hill Publishing, Auburn, NH, 2020).

______, 2020k, *The Seven Spaces of Creation: Octonion Cosmology* (Pingree Hill Publishing, Auburn, NH, 2020).

______, 2020l, *From Octonion Cosmology to the Unified SuperStandard Theory of Particles* (Pingree Hill Publishing, Auburn, NH, 2020).

______, 2021a, *Pioneering the Cosmos* (Pingree Hill Publishing, Auburn, NH, 2021).

______, 2021b, *Pioneering the Cosmos II* (Pingree Hill Publishing, Auburn, NH, 2021).

______, 2021c, *Beyond Octonion Cosmology* (Pingree Hill Publishing, Auburn, NH, 2021).

______, 2021d, *Universes are Particles* (Pingree Hill Publishing, Auburn, NH, 2021).

______, 2021e, *Octonion-like dna-based life, Universe expansion is decay, Emerging New Physics* (Pingree Hill Publishing, Auburn, NH, 2021).

______, 2021f, *The Science of Creation New Quantum Field Theory of Spaces* (Pingree Hill Publishing, Auburn, NH, 2021).

______, 2021g, *Quantum Space Theory With Application to Octonion Cosmology & Possibly To Fermionic Condensed Matter* (Pingree Hill Publishing, Auburn, NH, 2021).

______, 2021h, *21st Century Natural Philosophy of Octonion Cosmology , and Predestination, Fate, and Free Will* (Pingree Hill Publishing, Auburn, NH, 2021).

______, 2021i, *Beyond Octonion Cosmology II : Origin of the Quantum; A New Generalized Field Theory (GiFT); A Proof of the Spectrum of Universes; Atoms in Higher Universes* (Pingree Hill Publishing, Auburn, NH, 2021).

______, 2021j, *Integration of General Relativity and Quantum Theory: Octonion Cosmology, GiFT, Creation/Annihilation Spaces CASe, Reduction of Spaces to a Few Fermions and Symmetries in Fundamental Frames* (Pingree Hill Publishing, Auburn, NH, 2021).

______, 2022a, *New View of Octonion Cosmology Based on the Unification of General Relativit and Quantum Theory* (Pingree Hill Publishing, Auburn, NH, 2022).

______, 2022b, *The Gold Dust Beneath Hypercomplex Cosmology* (Pingree Hill Publishing, Auburn, NH, 2022).

Eddington, A. S., 1952, *The Mathematical Theory of Relativity* (Cambridge University Press, Cambridge, U.K., 1952).

Fant, Karl M., 2005, *Logically Determined Design: Clockless System Design With NULL Convention Logic* (John Wiley and Sons, Hoboken, NJ, 2005).

Feinberg, G. and Shapiro, R., 1980, *Life Beyond Earth: The Intelligent Earthlings Guide to Life in the Universe* (William Morrow and Company, New York, 1980).

Gelfand, I. M., Fomin, S. V., Silverman, R. A. (tr), 2000, *Calculus of Variations* (Dover Publications, Mineola, NY, 2000).

Giaquinta, M., Modica, G., Souchek, J., 1998, *Cartesian Coordinates in the Calculus of Variations* Volumes I and II (Springer-Verlag, New York, 1998).

Giaquinta, M., Hildebrandt, S., 1996, *Calculus of Variations* Volumes I and II (Springer-Verlag, New York, 1996).

Gradshteyn, I. S. and Ryzhik, I. M., 1965, *Table of Integrals, Series, and Products* (Academic Press, New York, 1965).

Heitler, W., 1954, *The Quantum Theory of Radiation* (Claendon Press, Oxford, UK, 1954).

Huang, Kerson, 1992, *Quarks, Leptons & Gauge Fields 2nd Edition* (World Scientific Publishing Company, Singapore, 1992).

Jost, J., Li-Jost, X., 1998, *Calculus of Variations* (Cambridge University Press, New York, 1998).

Kaku, Michio, 1993, *Quantum Field Theory*, (Oxford University Press, New York, 1993).

Kirk, G. S. and Raven, J. E., 1962, *The Presocratic Philosophers* (Cambridge University Press, New York, 1962).

Landau, L. D. and Lifshitz, E. M., 1987, *Fluid Mechanics 2nd Edition*, (Pergamon Press, Elmsford, NY, 1987).

Misner, C. W., Thorne, K. S., and Wheeler, J. A., 1973, *Gravitation* (W. H. Freeman, New York, 1973).

Rescher, N., 1967, *The Philosophy of Leibniz* (Prentice-Hall, Englewood Cliffs, NJ, 1967).

Rieffel, Eleanor and Polak, Wolfgang, 2014, *Quantum Computing* (MIT Press, Cambridge, MA, 2014).

Riesz, Frigyes and Sz.-Nagy, Béla, 1990, *Functional Analysis* (Dover Publications, New York, 1990).

Sagan, H., 1993, *Introduction to the Calculus of Variations* (Dover Publications, Mineola, NY, 1993).

Sakurai, J. J., 1964, *Invariance Principles and Elementary Particles* (Princeton University Press, Princeton, NJ, 1964).

Weinberg, S., 1972, *Gravitation and Cosmology* (John Wiley and Sons, New York, 1972).

Weinberg, S., 1995, *The Quantum Theory of Fields Volume I* (Cambridge University Press, New York, 1995).

INDEX

About the Author

Stephen Blaha is a well-known Physicist and Man of Letters with interests in Science, Society and civilization, the Arts, and Technology. He had an Alfred P. Sloan Foundation scholarship in college. He received his Ph.D. in Physics from Rockefeller University. He has served on the faculties of several major universities. He was also a Member of the Technical Staff at Bell Laboratories, a manager at the Boston Globe Newspaper, a Director at Wang Laboratories, and President of Blaha Software Inc. and of Janus Associates Inc. (NH).

Among other achievements he was a co-discoverer of the "r potential" for heavy quark binding developing the first (and still the only demonstrable) non-Aeolian gauge theory with an "r" potential; first suggested the existence of topological structures in superfluid He-3; first proposed Yang-Mills theories would appear in condensed matter phenomena with non-scalar order parameters; first developed a grammar-based formalism for quantum computers and applied it to elementary particle theories; first developed a new form of quantum field theory without divergences (thus solving a major 60 year old problem that enabled a unified theory of the Standard Model and Quantum Gravity without divergences to be developed); first developed a formulation of complex General Relativity based on analytic continuation from real space-time; first developed a generalized non-homogeneous Robertson-Walker metric that enabled a quantum theory of the Big Bang to be developed without singularities at t = 0; first generalized Cauchy's theorem and Gauss' theorem to complex, curved multi-dimensional spaces; received Honorable Mention in the Gravity Research Foundation Essay Competition in 1978; first developed a physically acceptable theory of faster-than-light particles; first derived a composition of extremums method in the Calculus of Variations; first quantitatively suggested that inflationary periods in the history of the universe were not needed; first proved Gödel's Theorem implies Nature must be quantum; provided a new alternative to the Higgs Mechanism, and Higgs particles, to generate masses; first showed how to resolve logical paradoxes including Gödel's Undecidability Theorem by developing Operator Logic and Quantum Operator Logic; first developed a quantitative harmonic oscillator-like model of the life cycle, and interactions, of civilizations; first showed how equations describing superorganisms also apply to civilizations. A recent book shows his theory applies successfully to the past 14 years of history and to *new* archaeological data on Andean and Mayan civilizations as well as Early Anatolian and Egyptian civilizations.

He first developed an axiomatic derivation of the form of The Standard Model from geometry – space-time properties – The Unified SuperStandard Model. It unifies all the known forces of Nature. It also has a Dark Matter sector that includes a Dark ElectroWeak sector with Dark doublets and Dark gauge interactions. It uses quantum coordinates to remove infinities that crop up in most

interacting quantum field theories and additionally to remove the infinities that appear in the Big Bang and generate inflationary growth of the universe. It shows gravity has a MOND-like form without sacrificing Newton's Laws. It relates the interactions of the MOND-like sector of gravity with the r-potential of Quark Confinement. The axioms of the theory lead to the question of their origin. We suggest in the preceding edition of this book it can be attributed to an entity with God-like properties. We explore these properties in "God Theory" and show they predict that the Cosmos exists forever although individual universes (or incarnations of our universe) "come and go." Several other important results emerge from God Theory such a functionally triune God. The Unified SuperStandard Theory has many other important parts described in the Current Edition of *The Unified SuperStandard Theory* and expanded in subsequent volumes.

Blaha has had a major impact on a succession of elementary particle theories: his Ph.D. thesis (1970), and papers, showed that quantum field theory calculations to all orders in ladder approximations could not give scaling deep inelastic electron-nucleon scattering. He later showed the eigenvalue equation for the fine structure constant α in Johnson-Baker-Willey QED had a zero at $\alpha = 1$ not 1/137 by solving the Schwinger-Dyson equations to all orders in an approximation that agreed with exact results to 4th order in α thus ending interest in this theory. In 1979 at Prof. Ken Johnson's (MIT) suggestion he calculated the proton-neutron mass difference in the MIT bag model and found the result had the wrong sign reducing interest in the bag model. These results all appear in Physical Review papers. In the 2000's he repeatedly pointed out the shortcomings of SuperString theory and showed that The Standard Model's form could be derived from space-time geometry by an extension of Lorentz transformations to faster than light transformations. This deeper space-time basis greatly increases the possibility that it is part of THE fundamental theory. Recently, Blaha showed that the Weak interactions differed significantly from the Strong, electromagnetic and gravitation interactions in important respects while these interactions had similar features, and suggested that ElectroWeak theory, which is essentially a glued union of the Weak interactions and Electromagnetism, possibly modulo unknown Higgs particle features, be replaced by a unified theory of the other interactions combined with a stand-alone Weak interaction theory. Blaha also showed that, if Charmonium calculations are taken seriously, the Strong interaction coupling constant is only a factor of five larger than the electromagnetic coupling constant, and thus Strong interaction perturbation theory would make sense and yield physically meaningful results.

In graduate school (1965-71) he wrote substantial papers in elementary particles and group theory: The Inelastic E- P Structure Functions in a Gluon Model. Phys. Lett. B40:501-502,1972; Deep-Inelastic E-P Structure Functions In A Ladder Model With Spin 1/2 Nucleons, Phys.Rev. D3:510-523,1971; Continuum Contributions To The Pion Radius, Phys. Rev. 178:2167-2169,1969; Character Analysis of U(N) and SU(N), J. Math. Phys. 10, 2156 (1969); and The Calculation of the Irreducible Characters of the Symmetric Group in Terms of the

Compound Characters, (Published as Blaha's Lemma in D. E. Knuth's book: *The Art of Computer Programming Vols. 1 – 4*).
In the early 1980's Blaha was also a pioneer in the development of UNIX for financial, scientific and Internet applications: benchmarked UNIX versions showing that block size was critical for UNIX performance, developing financial modeling software, starting database benchmarking comparison studies, developing Internet-like UNIX networking (1982) and developing a hybrid shell programming technique (1982) that was a precursor to the PERL programming language. He was also the manager of the AT&T ten-year future products development database. His work helped lead to commercial UNIX on computers such as Sun Micros, IBM AIX minis, and Apple computers.
In the 1980's he pioneered the development of PC Desktop Publishing on laser printers and was nominated for three "Awards for Technical Excellence" in 1987 by PC Magazine for PC software products that he designed and developed.
Recently he has developed a theory of Megaverses – actual universes of which our universe is one – with quantum particle-like properties based on the Wheeler-DeWitt equation of Quantum Gravity. He has developed a theory of a baryonic force, which had been conjectured many years ago, and estimated the strength of the force based on discrepancies in measurements of the gravitational constant G. This force, operative in D-dimensional space, can be used to escape from our universe in "uniships" which are the equivalent of the faster-than-light starships proposed in the author's earlier books. Thus travel to other universes, as well as to other stars is possible.
Blaha also considered the complexified Wheeler-DeWitt equation and showed that its limitation to real-valued coordinates and metrics generated a Cosmological Constant in the Einstein equations.
The author has also recently written a series of books on the serious problems of the United States and their solution as well as a book on the decline of Mankind that will follow from current social and genetic trends in Mankind.

In the past twenty years Dr. Blaha has written over 80 books on a wide range of topics. Some recent major works are: *From Asynchronous Logic to The Standard Model to Superflight to the Stars, All the Universe!, SuperCivilizations: Civilizations as Superorganisms, America's Future: an Islamic Surge, ISIS, al Qaeda, World Epidemics, Ukraine, Russia-China Pact, US Leadership Crisis, The Rises and Falls of Man – Destiny – 3000 AD: New Support for a Superorganism MACRO-THEORY of CIVILIZATIONS From CURRENT WORLD TRENDS and NEW Peruvian, Pre-Mayan, Mayan, Anatolian, and Early Egyptian Data, with a Projection to 3000 AD,* and *Mankind in Decline: Genetic Disasters, Human-Animal Hybrids, Overpopulation, Pollution, Global Warming, Food and Water Shortages, Desertification, Poverty, Rising Violence, Genocide, Epidemics, Wars, Leadership Failure.*

He has taught approximately 4,000 students in undergraduate, graduate, and postgraduate corporate education courses primarily in major universities, and large companies and government agencies.

He developed a quantum theory, The Unified SuperStandard Theory (UST), which describes elementary particles in detail without the difficulties of conventional quantum field theory. He found that the internal symmetries of this theory could be exactly derived from an octonion theory called QUeST. He further found that another octonion theory (UTMOST) describes the Megaverse. It can hold QUeST universes such as our own universe. It has an internal symmetry structure which is a superset of the QUeST internal symmetries.

Recently he developed Octonion Cosmology. He replaced it with HyperCosmos theory, which has significantly better features. He developed a fractionalization process for dimensions, particles and symmetry groups. He also described transformation that reduced particles and dimensions to a far more compact form. He also developed a precursor theory ProtoCosmos that leads to the HyperCosmos.

www.ingramcontent.com/pod-product-compliance
Lightning Source LLC
Chambersburg PA
CBHW082007190326
41458CB00010B/3102
9781737264057